CONCRETE TECHNOLOGY

CONCRETE TECHNOLOGY

THIRD EDITION

George R. White

Delmar Publishers Inc.®

NOTICE TO THE READER

Publisher does not warrant or guarantee any of the products described herein or perform any independent analysis in connection with any of the product information contained herein. Publisher does not assume, and expressly disclaims, any obligation to obtain and include information other than that provided to it by the manufacturer.

The reader is expressly warned to consider and adopt all safety precautions that might be indicated by the activities described herein and to avoid all potential hazards. By following the instructions contained herein, the reader willingly assumes all risks in connection with such instructions.

The publisher makes no representations or warranties of any kind, including but not limited to, the warranties of fitness for particular purpose or merchantability, nor are any such representations implied with respect to the material set forth herein, and the publisher takes no responsibility with respect to such material. The publisher shall not be liable for any special, consequential or exemplary damages resulting, in whole or in part, from the readers' use of, or reliance upon, this material.

Dedication

This text, CONCRETE TECHNOLOGY, is dedicated to the staff and research personnel of the Portland Cement Association who have been the leaders in most of the major developments of cement and concrete and related products in the United States. Their research work and construction application of this versatile construction material has opened new horizons for the future.

Cover credit: Photo by George R. White

Delmar Staff:

Senior Administrative Editor: Michael McDermott
Project Editor: Eleanor Isenhart
Production Coordinator: Wendy Troeger
Design Supervisor: Susan C. Mathews

For information, address Delmar Publishers Inc.
3 Columbia Circle, PO Box 15015
Albany, New York 12212-5015

Printed in the United States of America
Published simultaneously in Canada
by Nelson Canada
a Division of The Thomson Corporation

10 9 8 7 6

Library of Congress Cataloging-in-Publication Data

White, George R., 1923–
Concrete technology / George R. White. – 3rd ed., [rev.]
p. cm.
Includes bibliographical references and index.
ISBN 0-8273-3635-7
1. Concrete. I. Title.
TA439.W458 1991
666'.893–dc20 90-47267
CIP

CONTENTS

PREFACE

This is the third edition of the popular text CONCRETE TECHNOLOGY. Because of the dramatic changes in the construction field, references have been updated and others added to include innovative technical advancements. A section related to job searching and work opportunities has been included. All ASTM (American Society for Testing and Materials) specification references reflect the latest requirements. Also, included is a reference aid which shows the equivalent ASTM and AASHTO (American Association of State Highway and Transportation Officials) specifications. Our references would not be complete without mentioning the NICET (National Institute for Certification in Engineering Technologies). This certification program was designed for publicly-employed highway technicians involved in laboratory and field testing of materials.

Many exciting innovations in materials and construction procedures have appeared since the second edition of CONCRETE TECHNOLOGY. The author, in this edition, has drawn from the latest research findings and current building technology. The reader will find new information not only on changes in the development of cements, but also in the technology of construction. These changes have resulted in the rapidly increasing use of concrete to keep pace with construction progress. Among the newer topics covered in this revised edition are regulated set (expansive) cements, polymer concrete, wire and fiber reinforcement, and artificial aggregates. Most of the tables, charts, graphs, and other illustrative materials in this text have been revised to reflect the changes in specifications, especially as these changes show the elimination of bag measurement for concrete mixing and in its place the use of hundredweight. For convenience in calculating, a useful conversion table for the metric system is included in the appendix.

CONCRETE TECHNOLOGY is based on the fundamental principles of cement and concrete. The technical material is written in simple, easy-to-understand language. Students in vocational education and industrial arts will find this information of value in understanding cement, concrete, and concrete products as applied in the world of work. Individuals in apprenticeship training will learn the underlying reasons behind the "why" of cement and concrete. Students in post high school programs and advanced technical courses will find in CONCRETE TECHNOLOGY the basis on which they can build as they seek further understanding of more complex and detailed instructional materials available to them.

The new and updated information presented in this edition was prepared by George R. White. Mr. White has had a long association with all phases of cement and concrete technology through his work with educators, architects, engineers, contractors, and laboratory testing personnel. He is the author of several texts and has published many articles and technical presentations on the technology of concrete. Special mention is given to Clark V. White for his help and valuable assistance in reviewing the corrected manuscript for this third edition and for the new and updated information in Unit 9, Sampling and Testing Plastic Concrete. This will be a valuable aid to the readers of this text. I would also like to thank D.M. Weston of Bradford-Union Area Vo-Tech and Gary Edwards of SUNY–Delhi for their reviews of this edition.

This text material is based on the facts, tests, and authorities stated herein. It is intended for the use of professional personnel competent to evaluate the significance and limitations of the reported findings and who will accept responsibility for the application of the material it contains. The publisher, author, and resource personnel disclaim any and all responsibility for the accuracy of any of the sources on which this publication is based.

Photograph Credits:

The Portland Cement Association
George R. White

UNIT 1

JOB OPPORTUNITIES IN THE CONCRETE INDUSTRIES

The U.S. Department of Commerce[1] projects that the total labor force will be more than 127 million by 1995. Knowledge of specific industries can help future workers select the particular area of work in which they want to become involved. To assist in making the selection, those individuals who desire to enter the construction trades often raise several questions about this area of work. Is this field especially promising for employment? How many other individuals are entering the field? What type of training or education is necessary for employment? Since earnings vary in different occupations, prospective workers want to know about typical pay scales and working conditions. Other questions concern job advancement opportunities, personal recognition, and potential employers and their records.

The technology of most industries changes. As a result, some occupations are removed from the labor scene while other new and exciting occupations become available. Occupational skill requirements change as new industries are added to the economic picture. Most of the nation's workers are involved in producing services; therefore, future employment is expected to be greatest in this area. An exception is contract construction which may grow to a working force of about 7.9 million by the late 1990s.

THE CEMENT/CONCRETE INDUSTRIES

Since concrete is the most widely used building material in the world and its use is increasing, more trained personnel are needed in the cement and concrete construction industries and in the building and construction related occupations. These fields offer ample job opportunities with good wages, personal advancement, and safe, wholesome working conditions for laborers, technicians, apprentices in the trowel trades, inspectors, plant production and sales personnel, as well as for those with skills in other related areas.

THE CONSTRUCTION MARKET

All economic indicators point to a steady increase in the construction market over the next decade. Markets generally grow in proportion to the gross national product and the increase in population. An increased population means a larger expense for capital goods and the need for more homes, schools, shopping centers, roads, streets, and water and sewage systems. Fuel and water shortages will require more intense searches for quality supplies with the accompanying construction necessary to develop, manufacture, and transport these materials to demand centers.

More construction dollars must be spent to make the environment and working conditions better for all. As productivity increases so will leisure time and the requirement for facilities to make that time more enjoyable. Increased production of food and fiber will be necessary to support the increasing number of people. This means developing better

[1] U.S. Department of Commerce, (Washington, D.C.)

ways to grow and harvest crops, raise livestock, and produce the buildings and facilities necessary for these efforts.

The volume of construction over the next decade is expected to be the largest in the history of the United States. The continued growth of the construction industry will require even larger numbers of workers, technicians, scientists, engineers, and developers.

THE FUTURE

Not only will a course in concrete technology lead the student into rewarding jobs, but this basic course also serves as an introduction to advanced work in post secondary studies leading to an associate or baccalaureate degree in advanced technology or engineering.

Young people should carefully select the area in which they wish to make their careers. They should not automatically eliminate areas which do not seem to be growing at a rapid pace. Although growth is a key indicator of the future prosperity of an industry, many other jobs will be created by labor force specialties which have a rapid growth rate for necessary technicians.

Important new concerns include the management of solid and liquid waste created by our increasing population and life-cycle changes. Also, hazardous waste disposal or containment is of special interest to the future health and well-being of our society. Special concrete construction and unique concrete designs will play an important role in resolving many of our environmental challenges. This will mean more technical knowledge and specific job skills in different fields of work will be required.

Employers want people who are well trained and well educated. It is predicted that professional occupations (which require the most education) will show the fastest growth in the future. Training beyond high school has become a standard criteria in industry hiring. As more and more young people complete high school, there is greater competition in the work force for available jobs.

Building trades craftsmen represent the largest group of skilled workers in the nation's labor force. About 3 of every 10 skilled workers are employed in the building trades which comprise more than two dozen skilled trade groups. Concrete workers and cement finishers are classified as "Concrete Masons and Terrazzo Workers" by the U.S. Department of Labor, Bureau of Labor Statistics.

Building trades workers may be employed in factories, stores, mines, or other business establishments. Some workers are self employed and perform small contracting repair jobs for individual homeowners and others. Other workers are employed by construction industry contractors. Contractors are generally classified by the type of work they do: residential, commercial, highway, or industrial. Heavy construction contractors often take full responsibility for a complete job. They finish the work either with their own crews or through subcontractors. Trade contractors, such as a concrete contractor, usually complete only one portion of a large construction job.

To become proficient in a building trade, the student should progress from a formal learning environment into an apprenticeship training program for the trade involved. This on-the-job training period is designed to develop skills through continued class work, as well as actual job practice under the supervision of experienced craftsmen.

JOB OPPORTUNITIES

Because the concrete industries are expanding, job opportunities in concrete and concrete construction are plentiful. Despite many innovations in concrete construction, there is still no substitute for the services of the skilled worker.

Today's concrete worker is more advanced than his counterpart of 20 years ago. Workers in the past knew *how* to mix, handle, and place concrete; but they may not have known *why* certain practices were good and others bad. Today's worker must know how and why. The following paragraphs describe a few of the many opportunities for employment in the cement and concrete industries.

Ready-Mixed Concrete Industry

This industry, which is the largest consumer of portland cement, mixes and sells concrete for all types of construction. Employment opportunities include company owner, plant manager, laboratory and quality control technician, ready-mixed concrete truck driver, batch control operator, engineer, and sales representative. In many instances, the aggregates mining and manufacturing industry is combined with ready-mixed concrete production. This large industry offers many related job opportunities.

Precast Concrete Industry

Precast concrete is one of the fastest growing segments of the concrete industry. Increased quantities of precast concrete elements, such as wall panels, bridge beams, building components, roof and floor systems, decorative concrete pieces, and concrete pipe are being produced. Complete units such as buildings, septic tanks, water tanks, farmstead improvements, and bridges also are being made. The outlook for the use of precast concrete products continues to be favorable.

Employment opportunities include company owner, plant manager, engineer, batch control operator, laboratory and quality control technician, forming expert, reinforcing steel setter, concrete finisher, and salesperson.

Concrete Masonry and Concrete Products Industry

This industry uses mass-production methods to manufacture concrete units. The products include block, drain tile, pipe, roof tile, silo staves, manhole units, and telephone line conduit units. The employment opportunities include company owner, plant manager, batch control operator, laboratory and quality control technician, machine operator, and salesperson.

Construction Industry

The construction industry is very large. Since concrete is by far the most widely used of all construction materials, there are many opportunities for employment, including engineer, architect, contractor, construction superintendent, supervisor, estimator, batch control operator, laboratory and quality control technician, engineering technician, finisher, paving train operator, building inspector, job inspector, and highway, road or street department employee.

Allied Industries

There are many industries closely allied to the concrete industry. Included among these are product manufacturers, machinery and equipment manufacturers, accessory suppliers, form companies, finishing tool manufacturers, gravel and crushed rock suppliers, and special-aggregate producers.

All of the preceding industry categories require a knowledge of concrete. There are many others that cannot be listed because of space limitations. The United States Department of Labor Dictionary of Occupational Titles lists additional related occupations requiring a knowledge of concrete.

TYPE OF INDIVIDUALS WANTED

What type of person are the concrete industries seeking? Perhaps the most important personal qualification is integrity. Employers are looking for conscientious, hard working craftsmen. Other desirable qualifications include a cooperative and considerate attitude, initiative, and a readiness to assume responsibility.

A person who meets these qualifications can become part of an expanding business where job potential and opportunities are almost limitless.

EDUCATION AND TRAINING – THE KEY TO EMPLOYMENT

A high school diploma is a requirement for a majority of occupations. Many people who have less than twelve years of schooling may find their future job opportunities restricted to employment as private household workers and unskilled laborers. These areas do not represent expanding areas of employment.

Training and education provide the key that opens the door to employment. Through employment, the worker has an opportunity to gain experience and put his training to productive use. The combination of education, training, and employment experience is necessary if the worker is to do a job well and achieve the following:

- **Higher incomes.** Surveys show that trained individuals earn considerably more during their total years of employment than untrained and unskilled workers.
- **More job opportunities.** Unemployment is always most acute among the unskilled and untrained. Even when employment is high there is often a shortage of skilled workers.
- **Job stability.** A study in the northeastern states showed that a large percentage of vocational education graduates were working in the jobs related to their training.
- **Advancement.** A major advantage of working in concrete construction is that a skilled craftsman can advance to the position of foreman or become an owner of a contracting business.

Training for a job may take place at the beginning or in the later years of ones career. High schools, colleges, and universities (public and private) offer general as well as specialized job skills. Home study courses, city, county, and state governmental training programs, as well as the United States Armed Services, may give apprenticeship as well as semi-formal training. On-the-job training occurs in most occupations. Naturally, for each job classification there will be more than one way to reach your desired goal. The amount of training and developed skills will determine the level and pay scale as you enter that occupation.

General skills always have an effect on any specific job you acquire. So, too, do such attributes as personal health, good characteristics, dexterity, patience, accuracy, reliability, and acceptable work habits. All of these play a vital role in job acquisition.

For most concrete related jobs, you must have a high school diploma or its equivalent, be at least 18 years old, in good health and have a driver's license. Apprentices require on-the-job training in addition to 144 hours of classroom instruction. A written and physical exam is required. High school courses in blueprint reading, mathematics, and basic computer training are helpful.

STUDY/DISCUSSION QUESTIONS

1. Why must construction workers become highly skilled?
2. Describe the range of jobs available in the cement and concrete industries.
3. What construction groups influence the reputation of the concrete industry?
4. List the advantages of obtaining an education and job training.
5. Compare the present hourly wage rate in the construction trades with local service jobs.
6. What are the basic requirements for apprentice in the concrete trades?

UNIT 2

GETTING ACQUAINTED WITH CONCRETE

Concrete has several unusual characteristics that make it the most versatile and widely used of all construction materials. New developments resulting from years of research have provided today's concrete user with a unique, attractive, and practical product. Architects, engineers, and builders have used concrete with imagination and skill to create exciting and distinctive structures.

Freshly mixed concrete is a combination of *aggregates* (inert materials) and a paste composed of portland cement and water. The aggregates generally used are sand and gravel or crushed stone. These aggregates have no cementing value of their own; their function in concrete is to serve as a filler. The cement-water paste changes from a semifluid substance into a solid binder as a result of chemical reactions between the water and the various compounds in the cement. The final quality of the concrete depends upon the effectiveness of the hardened paste in binding the aggregate particles together and in filling the voids between the particles.

PROPERTIES DESIRED IN FRESH OR PLASTIC CONCRETE

Although freshly mixed concrete is plastic for a short time only, certain properties are desired at this stage because they affect the quality and cost of the hardened concrete. These properties, discussed in detail later, are defined as follows:

1. The wetness or dryness of the mix is referred to as the *consistency* or *slump.*
2. *Uniformity* indicates that the concrete is mixed thoroughly, has a standard appearance, and that all ingredients are evenly distributed in the mix.
3. *Workability* is the ease with which concrete is placed and consolidated.

PROPERTIES DESIRED IN HARDENED CONCRETE

Hardened concrete should be durable, strong, watertight, and resistant to abrasion. All of these properties are influenced by the quality of the portland cement paste. The relation between the amount of water and cement used in the paste is known as the *water-cement ratio.* The quality of the paste is determined by the total amount of water mixed with the cement. The paste can be compared to a powdered wood glue: mixed to the right consistency, it will glue. If too much water is added to the paste, it will be seriously weakened. It is for this reason that concrete is commonly proportioned according to the amount of water to be mixed with each hundredweight of cement. The quality of the hardened paste is governed by the water-cement ratio and the extent of the chemical reactions between the cement and the water.

Concrete must be durable enough to withstand extreme exposure conditions and to give long service with a minimum of upkeep. Concrete is used widely outdoors and is exposed to the destructive action of weathering. A durable concrete must be watertight. If water cannot penetrate the concrete to any appreciable extent, then repeated freeze-thaw cycles will cause little or no deterioration.

The most destructive weathering factor which can act on concrete is the freezing and thawing of concrete while it is wet or moist. Deterioration of the concrete may be caused by expansion of the paste, expansion of some of the particles of the aggregate, or by a combination of both. The introduction of entrained air into the concrete can help prevent this deterioration. Entrained air is the term applied to air bubbles of microscopic size in the concrete mixture.

Concrete must be strong enough to carry the heavy loads placed on it. Strength in concrete usually refers to compressive strength. While it often receives greater attention than either durability or watertightness, strength is the quality most easily controlled during construction. The strength of the concrete can be checked by tests of samples taken on the job. Fortunately, the same precautions which increase the strength of concrete also improve its durability and watertightness to about the same degree. It is possible, however, for concrete to be strong enough for its intended purpose and yet not be sufficiently durable and watertight for the conditions to which it is subjected.

If concrete is to be exposed to weather or other severe exposure conditions, it should be watertight. This requires a watertight paste. Tests show that a watertight paste depends on the water-cement ratio and the extent to which the concrete is cured or kept moist prior to exposure.

For certain uses, concrete must resist the abrasive action of wheeled vehicles, foot traffic, or flowing water. The same factors that increase durability, strength, and watertightness also increase the ability of concrete to resist abrasion or wear. When wear resistance is of primary importance, aggregate particles should be hard and tough. The aggregate should be as hard or harder than the cementing paste.

THE PROPERTIES OF CONCRETE

Concrete hardens in the presence of water. While most materials deteriorate when they become wet, concrete actually gains strength. This characteristic is very important when structures such as foundations and footings are to be placed in wet locations.

Concrete is composed of a variety of minerals and thus it does not corrode or decay. Concrete also resists vermin, termites, and rodents.

Freshly mixed concrete can be formed into practically any shape. Although many materials can be molded only when heated and require elaborate casting facilities, concrete can be molded at normal temperatures. This quality gives concrete a tremendous flexibility in construction applications.

Concrete will withstand heat and fire because it is made of nonflammable materials. For example, portland cement is manufactured at temperatures ranging from 2,600° F. to 3,000° F. Concrete has a high fire resistance which is a desirable quality in any construction, particularly in areas where fire protection is inadequate. However, a building constructed of concrete does not mean that the building cannot be damaged by fire. While concrete is fire resistant, it can lose strength when subjected to fire due to the fact that a high enough temperature will loosen the water from the concrete. In addition, the steel in reinforced concrete may lose strength, or the contents of a building may burn.

VARIABLES THAT INFLUENCE CONCRETE QUALITY

Water-Cement Ratio

The desirable properties of hardened concrete are influenced by the ratio of the amount of water used to the amount of cement used in the mix. This relationship is known as the water-cement ratio and is expressed in terms of the ratio of the weight of the water to the weight of the cement.

Aggregate Gradation

Concrete properties such as relative proportions, workability, economy, porosity, and shrinkage are all affected by the grading and maximum size of the aggregates used in the mix.

Variations in aggregate grading may have serious effects on the uniformity of concrete from one batch to another. Very fine sands are uneconomical; very coarse sands produce harsh, unworkable mixes. In general, aggregates that do not have a large deficiency or excess of any size produce the most satisfactory mixes.

Air Entrainment

The entrainment of air in concrete improves the resistance of the concrete to the adverse effects of freeze/thaw cycles. In addition, the ability of concrete to withstand the damaging effects of various deicing chemicals is improved. This resistance is particularly desirable in situations where the concrete is expected to be saturated with water.

When water freezes, it expands and may increase in volume as much as 9 percent over the original volume. Such an increase can cause pressures great enough to rupture the concrete and can cause scaling. Air-entrained concrete has microscopic air bubbles which act as reservoirs to relieve the pressure of excess water by absorbing it. These extremely small air voids protect the concrete and prevent damage during repeated freezing and thawing cyles.

Air entrainment has advantages in both plastic and hardened concrete:

PLASTIC CONCRETE

Air entrainment lowers the quantities of water and sand required per cubic yard.

Air-entrained concrete can be worked more easily.

Air entrainment tends to reduce aggregate segregation and bleeding.

Air-entrained concrete can be finished earlier than non-air-entrained concrete.

> Caution: Do not mistake the absence of bleeding water for the setting of the concrete and attempt to finish it too soon.

HARDENED CONCRETE

Air entrainment greatly improves the resistance of hardened concrete to the freeze/thaw cycle.

Air entrainment is effective in preventing serious surface scaling caused by deicing chemicals.

Air-entrained concrete is more watertight than non-air-entrained concrete.

CONSUMER PRODUCT SAFETY

Prolonged contact between wet cement and concrete mixtures should be avoided. To prevent such contact, it is advisable to wear protective clothing. Skin areas that have been exposed to wet cement or concrete, either directly or through saturated clothing, should be thoroughly washed with water. Contact with freshly mixed cement, mortar, grout, or concrete may cause skin irritation or burns. Wash skin areas which have been exposed to such contact promptly with water. If any cement or material containing cement gets into the eye, flush immediately with water and get prompt medical attention. Keep all cement and concrete mixtures and products out of the reach of small children.

STUDY/DISCUSSION QUESTIONS

1. What is concrete?
2. What are the inert materials?
3. What is the composition of the paste?
4. What properties are desired in fresh concrete?
5. Name four properties desired in hardened concrete.
6. Explain how the following variables influence the properties of concrete.
   a. Water-cement ratio
   b. Aggregate grading and maximum size
   c. Air entrainment
7. List the precautions to be taken with fresh concrete.

UNIT 3

USES OF CONCRETE

Plain concrete is defined as concrete containing no reinforcement. In addition to plain concrete, there are several other types of concrete which are used widely for construction.

REINFORCED CONCRETE

Reinforced concrete contains some form of reinforcement to increase its tensile strength. The assumption is made that the two materials (concrete and reinforcement) act together to resist forces.

The term *reinforcement* refers to the steel bars or welded wire fabric (wire mesh) placed in concrete to increase its tensile strength. While concrete is very strong in compression, that is, very strong for supporting loads that are placed directly upon it, it sometimes requires steel reinforcement to help it resist stresses or forces that tend to bend it or pull it apart.

PRESTRESSED CONCRETE

The technique of *prestressing* is generally applied to concrete members that are to be subjected to loads that cause bending. These members include beams, girders, and slabs. The concrete is compressed to counteract loads by stretching steel wires through the member. In reinforced concrete, all of the tensile stresses are carried by the reinforcing steel. In prestressed concrete, the entire cross section of the member is effective in supporting loads.

Prestressed concrete has many uses, including highway and railroad bridges, industrial buildings, precast joists, pressure pipe, poles, piling, water tanks, and roof systems. The savings in materials made possible by the prestressing process are impressive.

The prestressing force is applied by stretching the high-strength steel reinforcement. This stretching is accomplished either by pretensioning before the concrete is placed or by posttensioning after the concrete is hardened.

In *pretensioning*, steel wire is stretched in the form prior to the placement of the concrete. The concrete is then cast and allowed to harden, after which the wires are released from the outside anchorage. As the wires contract, they transmit compressive stresses to the concrete through the concrete-wire bond. Very high strength, stranded steel wires are used in pretensioning.

In the *posttensioning* process, steel cables are located in ducts embedded in the concrete. The steel rods or cables are mechanically anchored to one end of the beam. After the concrete has hardened, hydraulic jacks are used to stretch the steel rods. The free ends of the steel are then mechanically anchored. Large-diameter rods or cables are used. Since the mechanical anchor holds the tension, the concrete need not be bonded to the steel.

PRECAST CONCRETE

Precast concrete is a term applied to concrete objects cast and cured at a location other than that of their final use. Precast concrete is popular because of its economy and quality control aspects. When a large number of identical units are to be built, they may be precast on the job or at a specially equipped plant. The use of assembly-line production methods results in time and money savings, particularly in the forming process. Another advantage of precasting is that the work can be done during slack periods and stockpiled for later use. In this way, components such as beams, girders, and columns are available when needed.

Manufactured items such as concrete masonry, drain tile, concrete pipe, and silo staves are mass produced by large machines. While these items are precast, they are generally referred to as concrete products.

LIGHTWEIGHT CONCRETE

Conventional concrete weighs approximately 150 pounds per cubic foot. Lightweight concrete can be made with the use of either gas-generating chemicals, or lightweight aggregates such as expanded shale, clay, and slag. Concretes containing aggregates such as perlite and vermiculite are very light in weight and are used primarily as insulating materials.

Lightweight concrete may be classified according to the unit weight per cubic foot as follows:

1. *Insulating lightweight concrete.* Unit weight of about 20 to 70 pounds per cubic foot; compressive strengths seldom exceed 1,000 psi.
2. *Structural lightweight concrete.* Unit weight up to 115 pounds per cubic foot; 28-day compressive strength in excess of 2,000 psi.
3. *Semilightweight concrete.* Unit weight of 115 to 130 pounds per cubic foot. Sand of normal weight is substituted as a partial or complete replacement for the lightweight fine aggregate. Ultimate strength is comparable to that of normal weight concrete.

The first extensive use of lightweight concrete was for concrete block. Present uses of lightweight concrete include structural and insulation applications such as cast-in-place and precast walls, floors, and roof sections, and fireproofing.

The general principles of normal weight concrete proportioning may be applied to lightweight-aggregate concretes. When the aggregates are highly absorptive, however, some modifications are required.

The uniformity and quality of lightweight concrete depend primarily on the uniformity of the moisture content of the aggregate. If the moisture varies in the aggregates at the time of batching, then the slump, yield, and uniformity may be difficult to control. The moisture content of the aggregate should be known so that the aggregate batch weights can be adjusted to compensate for changes in absorbed water. If the final quantity of water added to the mix produces less than the required slump, additional water may be added to satisfy the slump requirements.

To control the uniformity of lightweight concrete, the cement content, slump, and volume of dry aggregates per cubic yard of concrete must be kept constant. This is accomplished by frequent unit weight tests of the fresh concrete.

THIN SHELLS

Thin shells of concrete were first used in Europe in the early 1920s. The use of concrete shells has increased greatly in the United States in recent years. As the term implies, the concrete is formed in thin sections which may be as thin as 2 1/2 inches. In many cases, even

Fig. 3-1 Shell action.

this thin cross section more than meets strength requirements. Rigid building codes, however, prescribe a minimum thickness of cover. As a result, the 2 1/2-in. shell thickness is usually the thinnest cross section permitted in construction in the United States. Shells of 5/8 in. thickness have been built outside the United States, however.

The strength and economy of shells is due to their shape. For example, if a sheet of paper is held along one end and then is lifted from the table, figure 3-1, it hangs limply from the support points. This sheet has practically no strength when cantilevered. If the sheet is now rolled into a semicircle and held along one edge, it not only cantilevers, but also supports the weight of paper clips.

If the sheet of paper is accordion pleated, its structural strength is increased in much the same manner as when it was curved. The depth of the roof and the intersection of the folds account for the great spanning and load-carrying abilities of folded-plate shells.

The savings in materials for the roof, impressive though they are, are only one of the many economies resulting from shell construction. Since considerable weight is eliminated in the roof, the size of the supporting columns and the amount of reinforcement can be reduced.

The economy of any shell roof is determined largely by the number of times the forms can be reused. In many cases, the forming costs can be reduced greatly by the use of mobile forms.

There are four types of commonly used shells: barrel, hyperbolic paraboloid, dome, and folded plate. Within each of these four categories, there are many variations in the shape of the shell.

There are two types of barrel shells: short and long barrels. Long barrels, figure 3-2, have short chord widths as compared to the span between the supporting ribs. Short barrels figure 3-3, have large chord widths as compared to the span between the ribs.

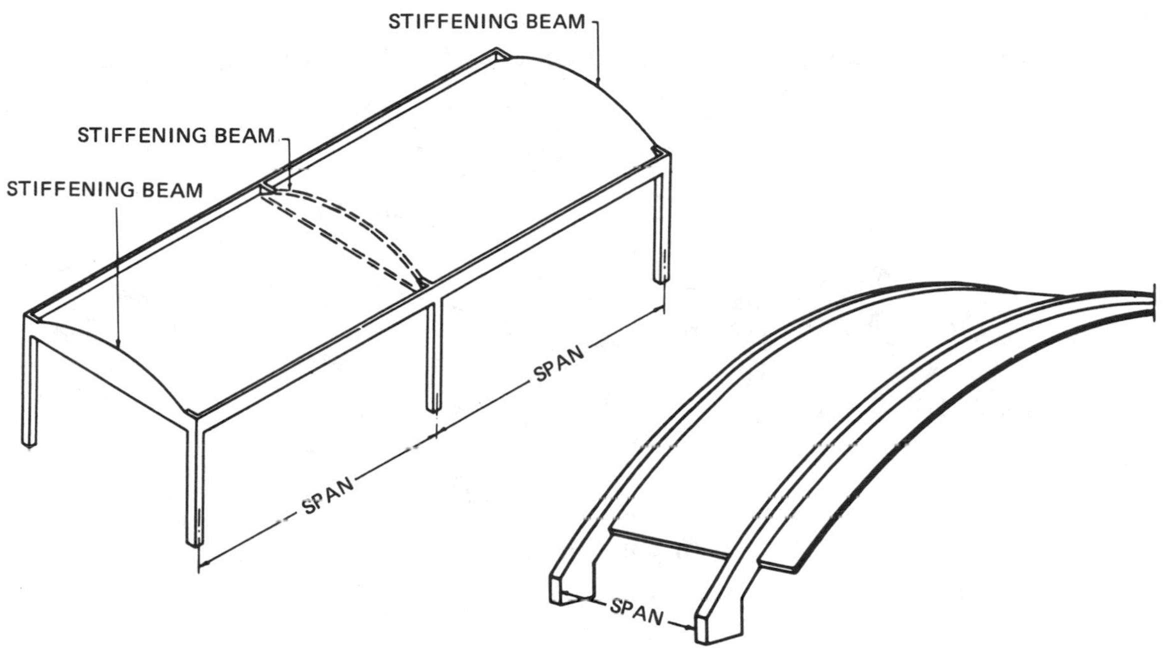

Fig. 3-2 Long barrel shell.

Fig. 3-3 Short barrel shell.

The hyperbolic paraboloid has a double curvature; that is, its surface is curved on two planes. The double curvature improves the stiffness of this type of shell and increases its ability to span and carry unsymmetrical loads. Two commonly used variations of the hyperbolic paraboloid shell are the saddle and umbrella shapes, figures 3-4 and 3-5. Despite the double curvature, the hyperbolic paraboloid is composed entirely of straight lines. As a result, it is possible to build forms using only straight lumber.

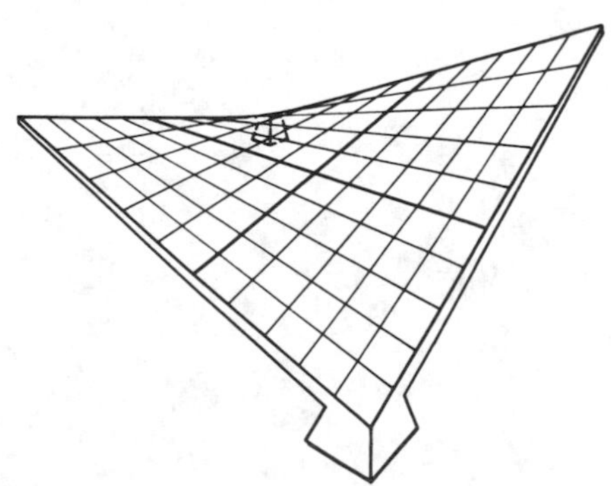

Fig. 3-4 Hyperbolic paraboloid, saddle type.

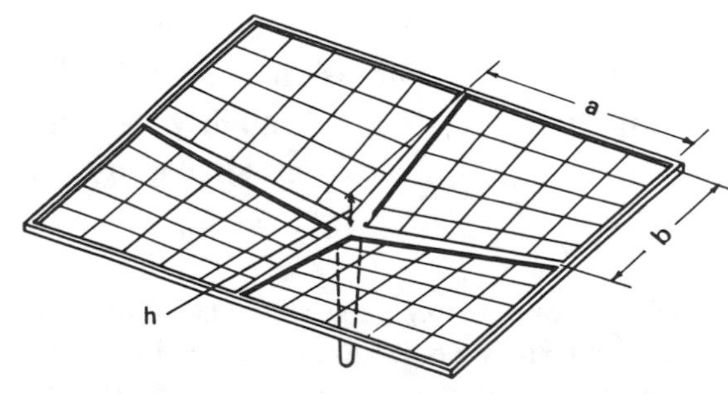

Fig. 3-5 Hyperbolic paraboloid, umbrella type.

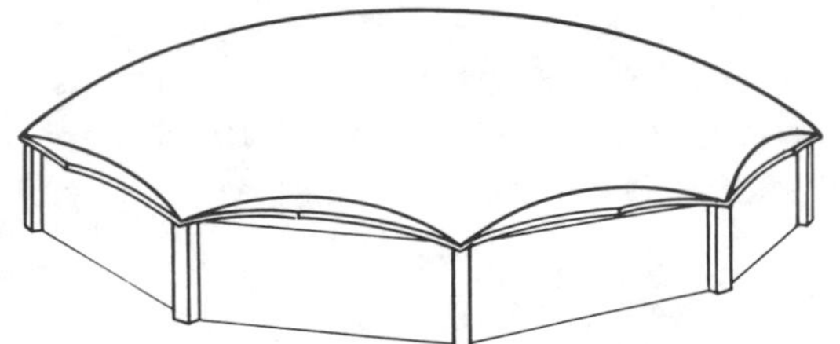

Fig. 3-6 Dome.

Dome shells are the aristocrats of roofs because of their perfectly symmetrical shape and the spacious, vaulted interiors they provide, figure 3-6.

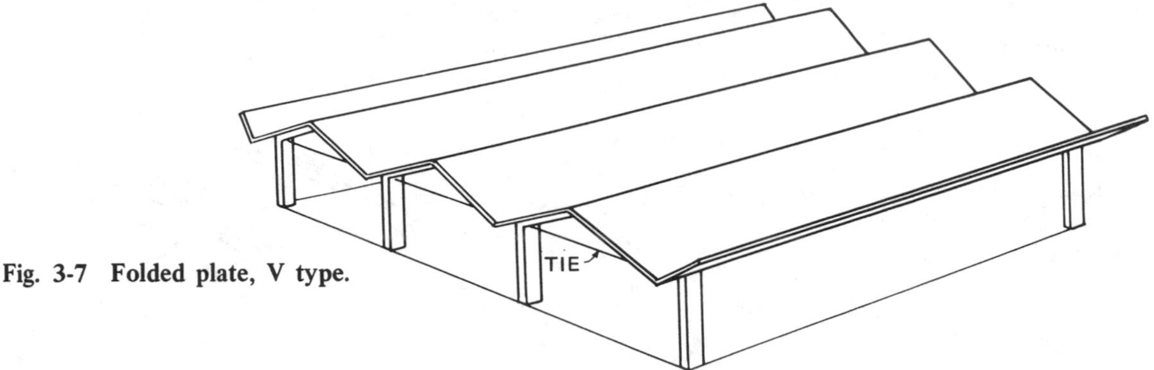

Fig. 3-7 Folded plate, V type.

There are three basic types of folded-plate shells: V-shaped, Z-shaped, and modified W-shaped. An example of the V-type shell is shown in figure 3-7.

PNEUMATICALLY APPLIED CONCRETE

Shotcrete is the term given to pneumatically applied portland cement plaster or concrete. A gun, operated by compressed air, applies the cement, water, and aggregate mixture, figure 3-8. The water may be added to the dry materials as they pass through the nozzle of the gun or it may be premixed with the materials.

When properly made and applied, shotcrete is an extremely strong and dense concrete and has high resistance to weathering. Its absorption rate is low, so that it is an excellent waterproofing medium. The resistance to abrasion of shotcrete is comparable to that of quality cast-in-place concrete made from similar aggregates.

Fig. 3-8 Gunning or pneumatic placing of concrete has many applications such as thin shell construction, slope paving, bank stabilization, and lining of swimming pools.

Fig. 3-9 Concrete panels, ranging in size from 8′ x 10′ x 4″ thick sections to 20′ x 21′ x 6″ thick sections (above) and larger, are used to enclose many farm and industrial buildings.

CONSTRUCTION WITH PREFABRICATED CONCRETE UNITS

Prefabricated or tilt-up construction is a fast, economical way to build single or multistory buildings, figure 3-9. The concrete walls are cast in sections or panels while laying flat on the floor or on a smooth bed of sand. After the wall panels are cured, they are tilted to their vertical positions and fastened together to make a wall. One way of fastening wall panels is accomplished by casting reinforced concrete columns between them. These reinforced columns are cast on top of previously cast pier footings and extend into the ground. These columns also may be cast on a continuous foundation. Reinforcing steel is continuous from the pier or foundation to the top of the column.

SANDWICH WALL

Sandwich wall panels, figure 3-10, are insulated concrete panels which are cast in a flat position. They may be precast or cast at the site. A typical 6-in. wall panel has a 1 1/2-in. section of expanded polystyrene insulation placed between two 2 1/4-in. sections of concrete. Sandwich wall panels are lifted into a vertical position and may be fastened together with columns which are cast in a manner similar to that for tilt-up construction.

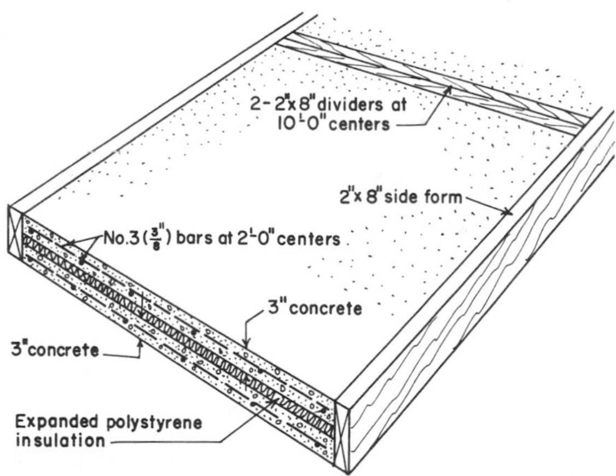

Fig. 3-10 The cross section of a typical insulated sandwich wall concrete panel. Construction of the panel consists of placing the bottom layer of concrete, then the insulation and reinforcing steel followed by the placing and finishing of the top layer of concrete.

Fig. 3-11 These concrete floors were cast on top of each other. Following curing, they were jacked into position. This method of erecting a floor system is called *lift slab construction.*

LIFT SLAB

Lift slab is a method of construction in which the floor slabs of a building are cast on the ground, one on top of another. The slabs are then jacked into place by hydraulic jacks and fastened to columns, figure 3-11. It is possible to construct buildings of several stories using this method.

STUDY/DISCUSSION QUESTIONS

1. Define reinforced concrete.
2. What is reinforcement and how is it used?
3. What is the difference between reinforced concrete and prestressed concrete?
4. What are the two general methods used for prestressing?
5. What materials savings are possible with prestressing?
6. Why is precast concrete popular?
7. Regular concrete weighs approximately 150 pounds per cubic foot. What is the weight limit for lightweight concrete?
8. Name the three types of lightweight concrete.
9. What are thin shells?
10. Name the four common types of shells.
11. Describe tilt-up construction.
12. What is pneumatically applied concrete?
13. Describe a concrete sandwich wall panel.
14. Complete this statement: "Lift slab is a method . . ."

UNIT 4

PORTLAND CEMENT

Portland cement is a finely ground material consisting primarily of compounds of lime, silica, alumina, and iron. When mixed with water, it forms a paste which hardens and binds the aggregates (such as sand, gravel, or crushed rock) to form a hard durable mass called *concrete.*

SETTING AND HARDENING

When portland cement is mixed with water, a paste is formed which first sets (becomes firm) and then hardens. The setting and hardening are due to the chemical reaction between the cement and water. This reaction is called *hydration* and results in a concrete which has a strong internal structure and is hard, durable, and watertight. The set and subsequent hardening process are the same whether the cement is used alone or in combination with aggregates. If the concrete is kept moist, the hydration reaction will continue for years and the concrete will become progressively stronger and more durable.

HISTORY OF CEMENTING MATERIALS

The history of cementing materials extends back to the time when prehistoric man abandoned his caves and started to build shelters. The first problem was to find a material to chink the stones to keep out the cold. The Assyrians and Babylonians used clay for this purpose. Then the Egyptians discovered lime and gypsum mortar which they used to build the pyramids. The Greeks made further improvements, and finally the Romans perfected a cement that produced structures of remarkable durability. The Colosseum and the great system of aqueducts and other cement-bonded structures built by the Romans are still in an excellent state of preservation.

Despite the early use of these materials, little was known of their chemistry. Substantial advances in the manufacture of cementing materials were not made from the time of the Romans until 1756. In that year, John Smeaton, who was building a lighthouse in the English Channel for the English government, discovered that when limestone and clay were mixed and burned, the mixture hardened into a solid mass under water as well as in air. Smeaton's discovery led to rapid improvements in cement and masonry construction.

In 1824, an Englishman, Joseph Aspdin, patented a process for the manufacture of an improved cement. The process consisted of heating a mixture of limestone and clay and then crushing the resulting product to a fine powder. He called this powder *portland* cement, because it produced a yellowish-gray concrete which resembled stone quarried on the Isle of Portland, England. Aspdin is generally recognized as the father of the modern portland cement industry.

The portland cement industry in North America began with the construction of a cement plant at Coplay, Pennsylvania, in 1872.

RAW MATERIALS

The raw materials for cement are divided into groups to indicate the component of the cement supplied by each group. The material groups are: the *calcareous* group supplying the lime component, the *siliceous* group supplying the silica component, the *argillaceous* group supplying the alumina component, and the *ferriferous* group supplying the iron component. Portland cement consists of a combination of the following compounds:

- Lime
- Silica
- Iron oxide
- Alumina
- Gypsum (added after burning)

Figure 4-1 illustrates typical sources of the components of portland cement.

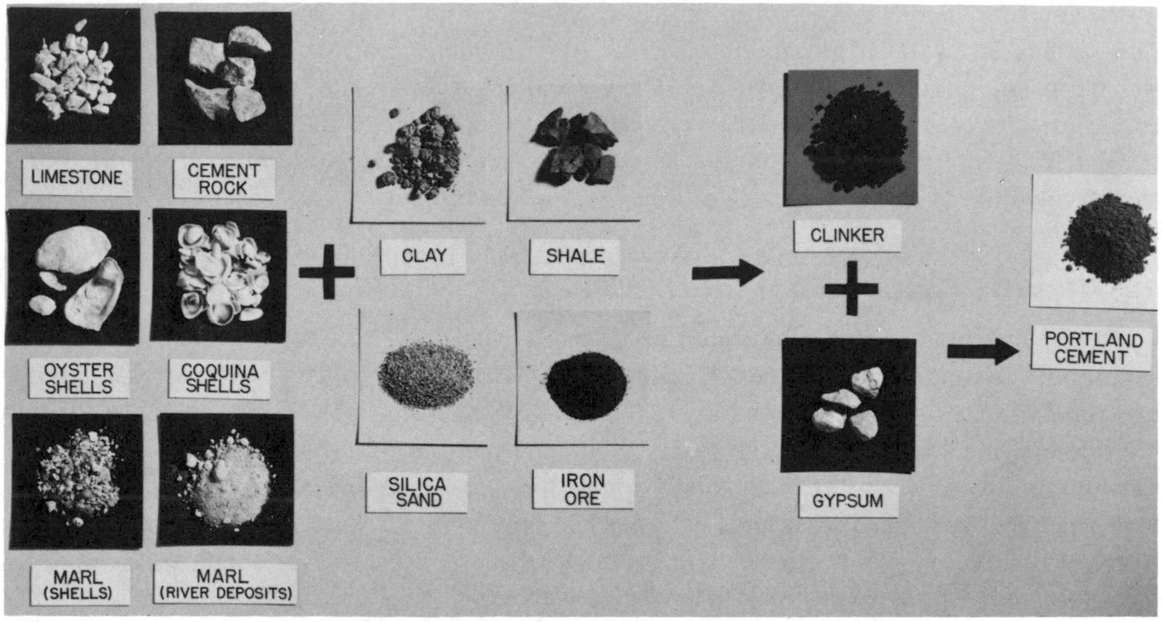

Fig. 4-1 Raw materials for portland cement.

The term *limestone* is used to include all carbonate rocks containing large quantities of calcium. In general, limestone is composed of calcium carbonate with varying minor percentages of magnesium carbonate, and clay and sand impurities. Clay and sand are not objectionable impurities when the rock is used for the manufacture of portland cement.

Cement rock is a low-magnesium limestone containing clay.

Marls are earthy, friable accumulations of calcareous materials secreted by plants and animals in lakes and marshes. Over long periods of time, the skeletal remains of plants mixed with animal shells can form beds as much as 30 feet thick. These beds contain substantial tonnages of material suitable for cement manufacture.

Shell marls consist of mixtures of fossil shells, shell fragments, and varying amounts of clay and sand.

Oyster and clam shells of recent origin can be collected and burned for lime. Adequate tonnages of shells for cement manufacture, however, cannot be obtained on the Atlantic coast of North America. In San Francisco Bay, shell deposits as much as 30 feet thick are found in brackish water areas. These deposits are suitable for cement manufacture. Along the Gulf of Mexico, oyster shells are the principal source of lime for cement manufacturers.

Coquina shells, used in Florida, are excavated under water and cleaned completely or partially of their sand by washing.

Clay and shale must be added when the limestone does not contain sufficient amounts of alumina and silica.

MANUFACTURING PROCESS

Portland cement is manufactured from carefully selected materials and by closely controlled processes. The first step is to blend the proper proportions of limestone, marl, or other calcareous materials with clay, shale, or blast-furnace slag. This mixture then is heated in a rotary kiln to a temperature of approximately 2,600° to 3,000° F. to form a clinker. During this burning operation, calcium oxide combines with the acidic components of the raw mix to form the following four compounds that make up most of the cement:

- Tricalcium silicate
- Dicalcium silicate
- Tricalcium aluminate
- Tetracalcium aluminoferrite

These compounds, in the presence of water, hydrate to form the hardened cement paste that binds the aggregate to make concrete. The clinker is cooled and pulverized. A small amount of gypsum is added to regulate the setting time. The pulverized product is portland cement. It is such a fine powder that nearly all of it will pass through a 200-mesh sieve (40,000 openings to the square inch). The portland cement is then stored in silos from which it is bagged or loaded for shipment.

PACKAGING AND STORAGE OF PORTLAND CEMENT

Much of the cement manufactured is shipped in bulk by rail or truck to ready-mix concrete producers and block manufacturers, or it is shipped directly to the construction site. The cement is unloaded by automatic equipment and placed into silos or hoppers for storage. Cement is also sold in paper bags. Each bag holds 1 cubic foot (cu. ft.) of cement and weighs 94 pounds. Cement is often measured in barrels and equals four bags or 376 pounds. The weight of a bag of masonry cement, averaging 70 pounds, is printed on the bag.

Portland cement will retain its cementing quality indefinitely if it does not come in contact with moisture. Thus, it should be stored in a location as dry and airtight as possible.

When bagged cement is stored for long periods, it sometimes acquires a "warehouse pack." This means that the cement is packed into a particular shape and will hold that shape in normal handling. This problem can usually be corrected by rolling the bag on the floor. When the cement is used, it should be free flowing and free of lumps. Cement containing lumps that cannot be broken up should not be used.

CLASSIFICATION OR TYPES OF PORTLAND CEMENT

Portland cement is a type of cement, not a brand name. Each cement manufacturer makes portland cement. All portland cements are hydraulic cements. This means that they will set and harden under water. Portland cements must meet one of the following ASTM (American Society for Testing and Materials) standard specifications:

C150 Portland Cement, Types I through V

C595 Blended Hydraulic Cements

- Portland blast-furnace slag cement—Type IS
- Portland-pozzolan cement—Type IP and P
- Slag Cement—Type S
- Pozzolan-modified portland cement—Type I(PM)
- Slag-modified portland cement—Type I (SM)

ASTM Designation C150

This specification covers eight types of portland cement. Type I, normal portland cement, is a general-purpose cement suitable for all uses not requiring the special properties of the other types of cement. Type I cement is used in pavement and sidewalk construction, reinforced concrete buildings and bridges, railway structures, tanks and reservoirs, culverts, water pipe, and masonry units. It is used for all cement or concrete construction not subject to sulfate attack from soil or water, or where the rise in temperature due to the heat generated by the hydration of the cement is not objectionable. Type IA is normal, air-entraining cement.

Type II cement, modified portland cement, has a lower heat of hydration than Type I cement and thus generates heat at a slower rate. It also has improved resistance to sulfate attack. It may be used to minimize the temperature rise in structures of considerable size such as large piers, heavy abutments, and heavy retaining walls. This is especially important when the concrete is placed in warm weather. In cold weather, when the heat generated is an advantage, Type I or III cement may be preferable. Type II cement is used when the added precaution against moderate sulfate attack is important, such as in drainage structures where the sulfate concentrations in the ground water are higher than normal but are not unusually severe. Type IIA is a moderate sulfate-resistant air-entraining cement.

Type III cement, high-early-strength portland cement, is used when high strengths are desired at very early periods, from one to three days after the concrete is placed. This cement is used when it is desired to remove the forms as soon as possible, or to put the concrete into service quickly. In addition, it is used in cold weather construction to reduce the necessary period of protection against low temperatures. Type III cement is also used when the high strengths desired at early periods can be obtained more satisfactorily or more economically with this cement than with richer mixes of Type I cement. Most prestressed concrete plants use this type of cement. Type IIIA is a high, early-strength, air-entraining cement.

Type IV, low-heat portland cement, is a special cement for use where the amount and rate of heat generated must be kept to a minimum. The development of strength in this cement is also at a slower rate. Type IV cement is used for those types of construction in which large masses of concrete are placed, such as for large gravity dams where the temperature rise resulting from the heat generated during hardening is a critical factor.

Type V, sulfate-resistant portland cement, is a special cement to be used only in construction exposed to severe sulfate action, such as in some western states where the soil or water may have a high alkali content. Type V cement has a slower rate of strength gain than normal portland cement.

Types IV and V cement normally are available only by special order.

Air-entraining cements. Air-entraining portland cements are designated as Types IA, IIA, and IIIA. These correspond to Types I, II, and III cements respectively, covered in ASTM C150 described previously. Small quantities of air-entraining materials are incorporated in Types IA, IIA, and IIIA cements by intergrinding them with the clinker during manufacture. These cements produce concrete that will resist severe frost action as well as the effects of chemicals applied for snow and ice removal. Concrete made with these cements contains tiny, well-distributed, completely separated air bubbles. The bubbles are so minute that there are billions of them in a cubic foot of air-entrained concrete.

White portland cement. White portland cement is manufactured to ASTM C150 specifications, usually Type I or Type III. White cement is made from raw materials selected to produce a pure white color. This cement is used primarily for architectural purposes such as precast curtain wall and facing panels, terrazzo surfaces, stucco, cement paint, tile grout, and decorative concrete. It is recommended that white portland cement be used wherever white or colored concrete or mortar is desired. White cement differs from gray cement only in its color and cost; white cement is more expensive than gray cement.

Portland Blast-furnace Slag Cement

Portland blast-furnace slag cement is divided into two types of cement: Type IS, portland blast-furnace slag cement, and Type IS-A, air-entraining portland blast-furnace slag cement. Granulated blast-furnace slag of selected quality is interground with portland cement clinker. These cements can be used for the same types of concrete applications as Type I or Type IA portland cement.

Portland-pozzolan Cement

Portland-pozzolan cement is also available in two types: Type IP, portland-pozzolan cement, and Type IP-A, air-entraining portland-pozzolan cement. In the manufacture of these cements, pozzolan (such as volcanic ash or volcanic rock) is blended with ground portland cement clinker. Types IP and IP-A portland-pozzolan cement are used largely for underwater structures such as bridge piers and dams.

Masonry Cements

Masonry cements conform to the requirements of ASTM C91. These cements are mixtures of portland cement, air-entraining additives, and supplemental materials selected for their ability to impart workability, plasticity, and water retention to masonry mortars.

Manufacturing controls permit the workability, strength, and color of masonry cements to be maintained at a uniform level. Masonry cements are classified as Type N, Type S, or Type M. In addition to mortar, masonry cements are used for parging and stucco. ASTM 270 covers *Specifications for Mortar in Unit Masonry.*

Special Cements

Special portland cements are types of portland cement some of which may not be covered by ASTM specifications.

Expansive cement (ASTM C845) is similar to portland cement; however, this cement has the unique property of expanding during the early hardening period after setting. At present, there are three types of expansive cement officially recognized. These cements are designated as Type EI, of which three varieties are recognized: EI (K); EI (M); and EI (S). Although these cements differ in composition, all three depend on the same chemical reaction to produce the expansion.

Most of the expansive cement sold in this country is used for shrinkage compensation. In other words, the cement is used to produce just enough expansion to counteract the shrinkage which occurs when the concrete dries. Modifications of the expansive cements have been used experimentally in attempts to produce a self-stressing concrete. Such a concrete would be capable of inducing prestress forces in embedded steel. Although some success in this application has been claimed by researchers in the Soviet Union, no large scale commercial use of the process has yet occurred in this country.

Regulated-set cement was first made in the Portland Cement Association laboratories in 1966 and the first commercial run was made in 1968. The setting time for this cement ranges from as little as five minutes to as long as one hour, depending on its formulation. This cement has achieved compressive strengths of 1,500 psi at two hours and 3,000 psi at four hours. The cement has the same composition as normal portland cements except for the addition of an active component which is responsible for the rapid strength development during the first few hours. After this period, the cement continues to harden in much the same way as a normal cement.

So far, commercial uses of regulated-set cement include roadway pavement repair, insulating roof deck concrete, and other special applications. The potential uses for this type of concrete are enormous, however. For example, in the precast concrete industry the fast-setting characteristics of this cement mean that multiple reuse of forms during a single work day can provide for greatly increased production rates with no additional investment in plant facilities.

A cement used to seal oil wells is called oil well cement. In general, this cement must be slow-setting and resistant to high temperatures and pressures. The American Petroleum Institute Specification for Oil-Well Cements (API Standard 10) covers nine classes of cements. Each class is used for a certain range of well depths. The petroleum industry also uses conventional types of portland cement in conjunction with suitable set-modifying admixtures.

Waterproofed portland cement is usually made by adding a substance to the portland cement clinker during the final grinding. This additive may be a small amount of calcium, aluminum, or other stearate. This cement is available in two colors, white and gray.

Plastic cements are made by adding plasticizing agents in amounts up to 12 percent of the total volume to Type I or II cement during the manufacturing process. Plastic cements are commonly used for making mortar, plaster, and stucco.

Cements made by combining water-reducing, retarding, and accelerating compounds must meet the ASTM C688 requirements.

AVAILABILITY OF CEMENTS

Some of the cements given in the previous section may not be available economically in all sections of the country. Before specifying the type of portland cement to be used, its availability should be determined. For example, Type V cement is used mainly in the western states and is seldom required in other areas.

Type I cement is generally carried by suppliers and is shipped to purchasers if the order does not specify the type of cement required. If a given type of cement is not available, comparable results frequently may be obtained with one of the available types. For example, high early-strength concrete can be made by using richer mixes (higher cement content) of Type I portland cement. In addition, the effects of the heat of hydration can be minimized by using smaller lifts or artificial cooling.

STUDY/DISCUSSION QUESTIONS

1. What is the difference between cement and concrete?
2. Explain the difference between concrete and mortar.
3. From what raw materials is portland cement made?
4. What is the volume of a bag of portland cement and how much does it weigh?
5. Explain the procedure for handling the condition known as warehouse pack.
6. Explain what is meant by hydration.
7. Is portland cement a brand or a type of cement?
8. What type of cement is most commonly used?
9. Are cements manufactured to meet particular specifications?
10. Which is the correct terminology: a cement sidewalk, or a concrete sidewalk?
11. List as many special cements as you can and describe their uses.

UNIT 5

MIXING WATER FOR CONCRETE

The principal reason for using water with cement is to cause hydration of the cement. Water added in excess of hydration requirements makes it possible to use more aggregates to produce a workable and economical mixture. Of the total volume of concrete, water comprises 21 percent (28 to 42 U.S. gallons per cubic yard).

There is a definite relation between the amount of water used and the quality of the resulting concrete. For example, too much water results in poor quality concrete. The amount of water must be limited to produce concrete of the quality required for the job. This relationship between the amount of water used and the quality of the concrete is covered in detail in later units of this text.

MIXING WATER FOR CONCRETE – HOW IMPURE CAN IT BE?

In general, any natural water that is drinkable and has no pronounced taste or odor is satisfactory for use as mixing water for making concrete. However, water suitable for mixing concrete may not be fit for drinking. To prevent possible health problems, many specifications for construction work state that the water used for the concrete must also be fit to drink. Tests should be made to insure that the setting time of the cement paste is not adversely affected by impurities in the mixing water. Excessive impurities may affect the setting time, concrete strength, and volume stability. In addition, the impurities may cause efflorescence, surface discoloration, and corrosion of steel. When water containing unknown impurities that could cause possible adverse effects is considered for use, particularly on larger concrete jobs, it is good practice to make a strength test. In this test, specimen cylinders made with concrete using the unknown water are compared with cylinders made with known acceptable water. Both 7-day and 28-day test cylinder strengths should be equal to the strengths of at least 90 percent of the specimens made with acceptable water.

Five typical analyses of city water supplies are shown in Table 5-1. The results of these

Analysis No.	1	2	3	4	5
Silica (SiO_2)	2.4	12.0	10.0	9.4	22.0
Iron (Fe)	0.1	0.0	0.1	0.2	0.1
Calcium (Ca)	5.8	36.0	92.0	96.0	3.0
Magnesium (Mg)	1.4	8.1	34.0	27.0	2,4
Sodium (Na)	1.7	6.5	8.2	183.0	215.0
Potassium (K)	0.7	1.2	1.4	18.0	9.8
Bicarbonate (HCO_3)	14.0	119.0	339.0	334.0	549.0
Sulfate (SO_4)	9.7	22.0	84.0	121.0	11.0
Chloride (Cl)	2.0	13.0	9.6	280.0	22.0
Nitrate (NO_3)	0.5	0.1	13.0	0.2	0.5
Total dissolved solids	31.0	165.0	434.0	983.0	564.0

Table 5-1 Typical analyses of city water supplies (parts per million).

analyses approximate the composition of the water supplies for about 45 percent of the cities in the United States with a population over 20,000. Water from any of these sources is suitable for making concrete. If a water supply is of unknown performance but is comparable in analysis to any of those given in the table, it is probably satisfactory for use in concrete.

Water containing less than 2,000 ppm (parts per million) of dissolved solids generally can be used for making concrete. Although concentrations exceeding 2,000 ppm are not always harmful, they can affect certain cements adversely. Where possible, high concentrations of dissolved solids should be avoided. Water departments usually will furnish an analysis of the local water.

HOW IMPURITIES AFFECT MIXING WATER

The following paragraphs summarize the effects of certain common impurities in mixing water on the quality of plain concrete.

Carbonates and bicarbonates of sodium and potassium have different effects on the setting time of different cements. Sodium carbonate may cause very rapid setting; bicarbonates may either accelerate or retard the set. In large concentrations, these salts can greatly reduce concrete strength. When the sum of these dissolved salts exceeds 1,000 ppm (0.1 percent), setting time tests and 28-day strength tests should be made.

If a naturally occurring water source has a high level of dissolved solids, these solids are usually sodium chloride or sodium sulfate. Both of these compounds can be tolerated in rather large quantities. Concentrations of 20,000 ppm of sodium chloride are generally acceptable. Mixing waters containing 10,000 ppm of sodium sulfate have been used satisfactorily.

Carbonates of calcium and magnesium are not very soluble in water. As a result, they do not affect the strength of concrete. Bicarbonates of calcium and magnesium are present in some municipal water supplies. Concentrations up to 400 ppm of bicarbonate ions are not considered harmful.

Magnesium sulfate and magnesium chloride can be present in concentrations up to 40,000 ppm without harmful effects on concrete strength. Calcium chloride can be used in concrete (except concrete containing aluminum or prestressing strands) in quantities up to 2 percent by weight of the cement. Calcium chloride is used to accelerate the hardening and the strength gain.

Natural groundwaters seldom contain more than 20 to 30 ppm of iron. However, acid mine waters may carry rather large quantities of iron. Iron salts in concentrations up to 40,000 ppm usually do not affect mortar strengths adversely.

Salts of manganese, tin, zinc, copper, and lead in mixing water may cause a significant reduction in strength and cause large variations in setting time. Zinc, copper, and lead salts are the most active. Other salts that are especially active as retarders include sodium iodate, sodium phosphate, sodium arsenate, and sodium borate. All of these salts can greatly retard the set and strength development when present in concentrations of a few tenths of a percent by weight of cement. Concentrations of these salts up to 500 ppm generally can be tolerated in the mixing water. Sodium sulfide is another salt which may be detrimental to concrete. Concrete made from water containing as little as 100 ppm of sodium sulfide should be tested.

Seawater, which contains salts in quantities up to 35,000 ppm (3.5 percent), is generally suitable for use as mixing water for nonreinforced concrete. Although concrete made with seawater may have higher early strength than normal concrete, strengths at later ages (after 28 days) may be lower. This strength reduction can be prevented by reducing the water-cement ratio. Quality concrete can be made with seawater if the mix is properly adjusted.

If suitable fresh water is not available, seawater may be used for making reinforced concrete. The use of seawater may increase the risk of corrosion; however, this risk is reduced if the reinforcement has sufficient cover and if the concrete is watertight and contains an adequate amount of entrained air. Reinforced concrete structures made with seawater and exposed to a marine environment should have a water-cement ratio of less than 5 gallons of seawater per bag; the reinforcement cover for this concrete should be at least 3 inches.

Seawater should not be used for making prestressed concrete in which the prestressing steel is in contact with the concrete.

Sand and gravel removed from seawater are sometimes used in the making of concrete. The amount of sea salt on the aggregate usually is not more than about 1 percent of the weight of the mixing water. These aggregates combined with drinkable water contributes less salt to the mix than does seawater.

The decision to use acidic mixing water should be based on the concentration (in parts per million) of the acids in the water. On occasion, use of the water is based on the pH of the water; pH is a measure of the hydrogen ion concentration. The pH of neutral water is 7.0; values below 7.0 indicate acidity and values above 7.0 indicate alkalinity. The pH value is an intensity index and is not the best measure of potential acidic or alkaline (basic) reactions.

Mixing waters which contain hydrochloric, sulfuric, and other common inorganic acids in concentrations as high as 10,000 ppm generally have no adverse effect on concrete strength. Acidic water with a pH value less than 3.0 may create handling problems and should be avoided.

Mixing water containing sodium hydroxide concentrations of 0.5 percent by weight of cement do not greatly affect the concrete strength provided quick set is not induced. Higher concentrations of sodium hydroxide, however, may reduce the strength.

Potassium hydroxide in concentrations up to 1.2 percent by weight of cement has little effect on the strength developed by some cements. However, the same concentrations of potassium hydroxide when used with other cements may substantially reduce the 28-day strength.

Most water containing industrial waste has less than 4,000 ppm of total dissolved solids. When such water is used as mixing water in concrete, the reduction in compressive strength generally is not greater than about 10 percent. Waste waters from industries such as tanneries, paint factories, coke plants, and chemical and galvanizing plants may contain harmful impurities. It is recommended that tests be performed on any waste water containing even a few hundred parts per million of unusual solids.

Typical sewage may contain about 400 ppm of organic matter. After the sewage is diluted in a good disposal system, the organic concentration is reduced to about 20 ppm or less. This amount is too low to have any significant effect on concrete strength.

Sugar in amounts as small as 0.03 to 0.15 percent by weight of cement usually retards the setting of cement. The upper limit of this range varies with different cements. The 7-day strength may be reduced and the 28-day strength may be improved. When the amount of sugar is increased to about 0.20 percent by weight of cement, the set is usually accelerated. Sugar in quanitites of 0.25 percent or more by weight of cement may cause rapid setting and a substantial reduction in 28-day strength.

If the amount of sugar in the mix water is less than 500 ppm, it generally has no adverse effect on the concrete strength. If the concentration exceeds this amount, however, tests for setting time and concrete strength should be made.

Suspended clay or fine rock particles in quantities up to 2,000 ppm can be tolerated in the mixing water. Higher amounts may not affect the strength but may influence other properties of some concrete mixtures. Muddy water should remain in settling basins before used to reduce the amount of silt and clay added to the mix.

Various kinds of oil may be present in the mixing water. Mineral (petroleum) oil alone (not mixed with animal or vegetable oils) probably has less effect on strength development than other oils. However, mineral oil in concentrations greater than 2 percent by weight of cement may reduce the concrete strength by more than 20 percent.

The presence of algae in the mixing water may cause an excessive reduction in strength either by combining with the cement to reduce the bond or by causing a large amount of air to be entrained in the concrete. Algae may be present on the aggregates as well, resulting in a reduction of the bond between the aggregate and the cement paste.

STUDY/DISCUSSION QUESTIONS

1. What is the purpose of mixing water?
2. Why is more water used than is required to achieve its principal purpose?
3. What is a good rule for determining whether water is satisfactory as mixing water for making concrete?
4. If water of unknown performance is to be used for concrete, explain how to determine whether it will be acceptable.
5. What are some of the effects impurities in the mixing water may have on concrete?
6. Should seawater be used in concrete? Why?
7. After what process is sewage water usable for making concrete?
8. What effect does algáe have if it is present in the mixing water?

UNIT 6

AGGREGATES FOR CONCRETE

Sand, gravel, and crushed stone are the aggregates most commonly used in concrete to provide volume at low cost. Since aggregates make up about 60 to 80 percent of the volume of concrete, they can be called a filler material.

The selection of the aggregates to be used is very important in making concrete. The characteristics of the aggregates greatly influence the properties, mix proportions, and economy of the concrete. The aggregates should consist of particles having adequate strength and resistance to exposure conditions. They should not contain materials having harmful effects such as dirt, clay, coal, or organic matter. They should be graded in size to achieve the best economy from the paste. Aggregates are generally classified as fine and coarse.

CHARACTERISTICS OF AGGREGATES

Basically, good aggregates must have certain characteristics if the resulting concrete is to be workable, strong, durable, and economical. These characteristics include abrasion resistance, resistance to freezing and thawing, compressive strength, chemical stability, and good particle shape and surface texture.

The abrasion resistance of an aggregate is often used as a general index of aggregate quality. Abrasion resistance is essential when the aggregate is used in concrete subject to abrasion, as in floors and pavements.

Resistance to freezing and thawing is important when the aggregates are used in exposed concrete. The freeze-thaw resistance of an aggregate is related to its porosity, absorption, and pore structure. If an aggregate particle absorbs so much water that insufficient pore space is available to accommodate the water expansion that occurs during freezing, the concrete may be vulnerable to freezing.

The compressive strength of an aggregate is its resistance to compressive forces.

The chemical stability of an aggregate indicates that the aggregate will not react chemically with cement or be affected chemically by other external influences. In some areas, aggregates with certain chemical constituents react with alkalies in cement. This alkali-aggregate reaction may cause abnormal expansion and map-cracking of the concrete.

The particle shape and surface texture of an aggregate influence the properties of fresh concrete more than they affect the properties of hardened concrete. Particles with a rough texture or flat and elongated particles require more cement paste to produce workable concrete than is required by rounded or cubical aggregates. However, crushed and uncrushed aggregates generally give essentially the same strength for the same cement content.

Coarse aggregate particles should be cubical in shape and free of excessive amounts of flat and elongated pieces. Long slivers of aggregate pieces should be avoided or at least limited to 15 percent by weight of the total aggregate. This requirement is equally important for fine aggregate, since sand made by crushing stone often consists of angular particles. When manufactured sand is used for fine aggregate in concrete, care should be taken to avoid materials having an abundance of thin, sliver-type particles. This is particularly

CAUTION: Sodium hydroxide should not be handled with moist hands as serious burns may result. Do not spill the solution, as it is highly injurious to clothing, leather, and most other materials.

As soon as the solution of sodium hydroxide is added to the sand, shake the contents of the bottle thoroughly and then allow the bottle to stand for 24 hours. The color of the liquid will indicate if the sand contains vegetable matter. A colorless liquid indicates clean sand free from vegetable matter. A straw-colored solution indicates some vegetable matter but not enough to be objectionable. Darker colors mean that the sand contains harmful amounts of organic materials and should not be used unless it is washed and tested again.

AGGREGATE SIZE

Fine Aggregate

Fine aggregate consists of particles 1/4 inch and less in size. Natural and manufactured sands are the most common examples of fine aggregate. They have particles ranging in size from 1/4 inch to those small enough to pass through a sieve having 100 openings to the inch.

Coarse Aggregate

Coarse aggregate is usually gravel or crushed stone. Sizes range from 1/4 inch up to the maximum size permitted for the job.

Bank-Run and Commercially Recombined Aggregates

Bank-run material is aggregate that is used as taken from the quarry or gravel pit. The natural mixture of fine and coarse aggregates as taken from a gravel bank or crusher usually does not make the most economical concrete unless it is first screened to separate the fine material from the coarse and then is recombined in the correct proportions. Most gravel banks contain too much sand in proportion to the coarse material. More cement paste is required to produce concrete of a given quality when there is a high proportion of fine aggregate.

When using bank-run material, it is generally advisable to screen the material to separate it into fine and coarse aggregates. Some commercial firms also sell a mixed aggregate for which the sand and gravel are separated and then recombined into the correct proportions for concrete.

Bank-run aggregates frequently are not clean enough for use until they are washed to remove silt, clay, or other materials detrimental to good concrete. Since clean aggregates are essential to quality concrete, thorough washing is required.

Concrete aggregates are sold by weight in many localities. For estimating purposes, it can be assumed that a ton of aggregate contains approximately 20 to 22 cubic feet of sand crushed stone, or gravel.

ARTIFICIAL AGGREGATES

Artificial aggregates have been used in concrete for some time. Slag from was used during the last half of the 19th century. The Hayde process for exp shale into lightweight aggregates was developed in 1917. Slag and expande still the major synthetic aggregates in use. However, other materials are c

true if the amount of water required to make a cubic yard of concrete is considerably greater than if well-graded aggregate particles are used.

Hard and durable aggregates are best suited for use in concrete. Aggregates that are soft and flaky and show rapid wear when exposed to the weather generally are unsatisfactory. Weak, friable, or laminated aggregate particles are undesirable. Shale, stones laminated with shale, and most cherts are to be avoided as aggregates.

CLEAN AGGREGATES DESIRABLE

The best aggregates are clean and free of fine dust, loam, silt, clay, or vegetable matter. These materials are objectionable because they prevent the cement paste from binding the aggregate particles. The strength of the concrete is reduced as a result. Concrete made with dirty aggregates hardens slowly, or it may never harden enough to serve its intended use.

TESTING AGGREGATES FOR QUALITY

If it is necessary to use aggregates of unknown quality, they should be carefully examined and tested to insure that they are suitable for making concrete.

The silt test is used to detect the presence of extremely fine material. The colorimetric test is used to detect the presence of harmful amounts of vegetable matter.

The Silt Test

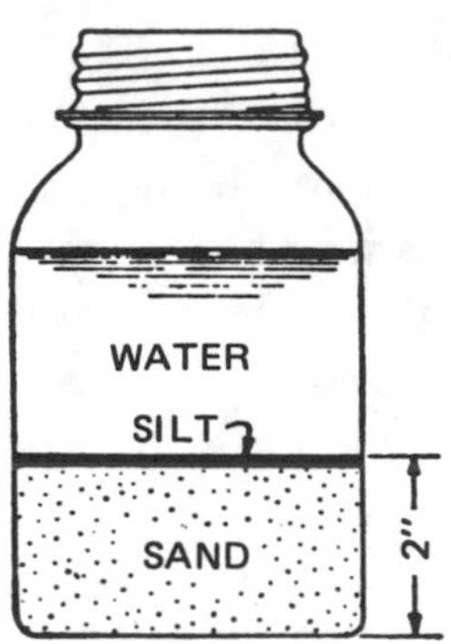

Fig. 6-1 Silt test.

To perform the silt test, use an ordinary quart milk bottle or quart fruit jar, figure 6-1. Fill the container to a depth of two inches with a sample of the dry sand to be tested. Add water until the bottle or jar is about three-fourths full. Shake the bottle vigorously for one minute. Then shake the bottle a few times in a sideways direction to level the sand. Allow the jar to stand for an hour. During this time any silt present will be deposited in a layer above the sand. If the layer is more than 1/8 inch thick, the sand from which the sample was taken is not satisfactory for concrete work unless the excess silt is removed. Excess silt can be removed by washing the sand.

The Colorimetric Test

The colorimetric test requires an ordinary 12-ounce prescription bottle, such as druggists or physicians use. The bottle is filled to the 4 1/2-ounce mark with a sample of the sand, figure 6-2. A 3 percent solution of caustic soda is added to the bottle. This solution is made by dissolving 1 ounce of sodium hydroxide (household lye) in a quart of water, preferably distilled. The solution should be kept in a glass bottle tightly closed with a rubber stopper.

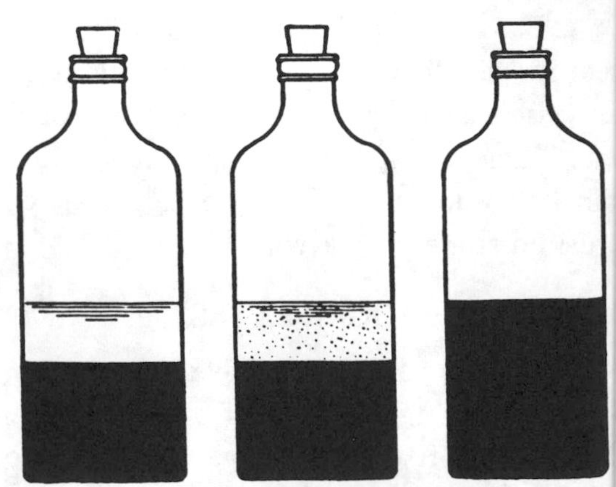

Fig. 6-2 Colorimetric test.

developed because of rapidly dwindling supplies of natural sands and gravels. In addition, new applications of concrete require materials with a wider range of unit weights.

Ultra-lightweight materials are used for insulation and for concretes which can be sawed or nailed. These materials include perlite, vermiculite, expanded glass, ceramic sphères, carbonized cereals, and expanded plastics. All of these materials have unit weights ranging from 4 to 30 pounds per cubic foot.

Lightweight aggregates, with unit weights of 40 to 70 pounds per cubic foot, are used primarily for reducing the dead loads of structures. These materials may also be used for their insulating properties. Lightweight aggregates include expanded clay, shale, slate, sintered fly ash, expanded slag, furnace clinker, fired urban waste, crushed bricks, and petroleum coke.

Aggregates ranging in weight from 70 to 110 pounds per cubic foot, include iron slag, glass frits, calcined bauxite, and compacted cement. Research is in process on the production of stable normal weight aggregates in portland cement kilns using cement raw materials.

Heavyweight aggregates having unit weights in excess of 120 pounds per cubic foot include steel scrap and steel shot. These materials may be replaced by compounds containing iron and phosphorus.

Lightweight Aggregates

The most commonly used aggregates, such as sand, gravel, crushed stone, and air-cooled blast-furnace slag, produce normal weight concrete. This is concrete with weights ranging from about 140 to 160 pounds per cubic foot. Expanded shale, clay, slate, and slag are used as aggregates to produce structural lightweight concretes having unit weights ranging from about 85 to 115 pounds per cubic foot. Other lightweight materials, such as cinders, pumice, scoria, perlite, vermiculite, and diatomite, are used to produce insulating concretes weighing about 20 to 70 pounds per cubic foot.

Heavyweight Aggregates

Materials such as barytes, limonite, magnetite, ilmenite, iron, and steel particles are used to produce heavyweight concrete. This type of concrete is used primarily for shielding against nuclear radiation.

GRADATION OF AGGREGATES

Figures 6-3 through 6-6 show the appearance of aggregates before and after grading into various sizes.

To make good concrete, each aggregate particle, regardless of size, must be completely surrounded by the cement paste. It is important that fine and coarse aggregates be proportioned so that the finer particles will fill the spaces between the larger particles. This results in the most economical use of cement paste to fill the voids and bind the aggregate particles together.

A simple demonstration of this concept can be made. Obtain three transparent containers of equal capacity and place the same amount of water in each. This water will represent a selected quality of cement paste. Fill each container approximately one-fourth full. Now add as much coarse aggregate to one container as the cement paste (water) will cover. Do the previous step with fine aggregate in the second container. Add either a com-

Fig. 6-3 Coarse aggregate before being separated. Note how the smaller pieces fit among the larger ones in the mixed aggregate.

Fig. 6-4 Well-graded coarse aggregate after being separated into three sizes: (a) 1/4 to 3/8 in., (b) 3/8 to 3/4 in., (c) 3/4 to 1 1/2 in.

Fig. 6-5 Sample of well-graded sand before and after separation into various sizes. Particles vary from dust to those which will just pass through a No. 4 sieve (approximately 1/4 in.).

Fig. 6-6 Sample of sand which lacks particles larger than 1/16 in. (top). Sample is separated into four sizes (bottom). More cement is required when sand is fine. This is not a good concrete sand.

bined aggregate or the proper gradation of aggregates to the third container. Note that the same amount of cement paste (water) covers the surface area of a greater volume of material in the last container. This shows that better economy is obtained by the use of the proper gradation of aggregates. The demonstration shows that the cement paste requirement for concrete is proportional to the void content of the combined aggregates. It can be seen that the amount of concrete (of a given quality) that can be obtained from a bag of cement depends on the filler material.

DETERMINING MAXIMUM SIZE OF COARSE AGGREGATE

As stated previously, less paste is necessary with an increase in the coarse aggregate size. As a result, more concrete is obtained per bag of cement. Generally, this means that the most economical mix is obtained by using the largest coarse aggregate practical for the job. The maximum size of aggregate used depends on the size and shape of the concrete members and the amount and distribution of the reinforcing steel.

The maximum size of aggregate should not exceed the following guidelines:

1. One-fifth the dimension of nonreinforced members,
2. Three-fourths the clear spacing between reinforcing bars or between reinforcing bars and forms, or
3. One-third the depth of nonreinforced slabs on grade.

The common maximum aggregate sizes are 3/8, 1/2, 3/4, 1, 1 1/2, and 2 inches.

HANDLING AND STORING AGGREGATES

Aggregates should be handled and stored to prevent segregation of sizes and contamination with harmful materials. Stockpiles should be built up in layers of uniform thickness. Cone-shaped piles should not be used as this method of storing results in segregation of sizes. Damp fine aggregate has less tendency to segregate than dry material. Aggregates should be removed from stockpiles in approximately horizontal layers to minimize segregation.

STUDY/DISCUSSION QUESTIONS

1. Why are aggregates referred to as filler material?
2. What are the two classifications of aggregates?
3. What are some of the characteristics of a good aggregate?
4. What types of aggregates should be avoided?
5. What effect will dirty aggregates have on the strength of concrete?
6. How is sand tested for silt and organic matter?
7. What is a colorimetric test and how is it conducted?
8. Define the terms fine aggregate and coarse aggregate.
9. Why is bank-run gravel seldom suitable for concrete work?
10. How can bank-run gravel be made usable?
11. Approximately how many cubic feet are contained in a ton of sand? Crushed stone? Gravel?
12. Why are lightweight aggregates used for certain jobs?
13. Why does the proportion of fine and coarse aggregates affect the concrete yield (the amount of concrete that can be obtained from a bag of cement and a fixed proportion of water)?
14. Explain the term *void content* as it applies to concrete aggregates.
15. What factors determine the maximum size of aggregate that can be used?
16. How should aggregates be stockpiled on the job?

UNIT 7

AIR-ENTRAINED CONCRETE

One of the greatest advances in concrete technology was the introduction of air-entrained concrete in the mid 1930s. Today, the use of air-entrained concrete is recommended for nearly all construction requirements.

The main reason for using intentionally entrained air is to improve the resistance of concrete to freezing and thawing. However, there are other important benefits due to entrained air in both fresh and hardened concrete.

Air-entrained concrete is produced by using an air-entraining cement or an air-entraining admixture during the mixing of the concrete. Unlike trapped air voids, intentionally entrained air bubbles are extremely small; the diameters of these bubbles range from about a thousandth of an inch to about a hundredth of an inch. These bubbles are not interconnected and are well distributed throughout the paste (water + cement + air). As many as 400 to 600 billion bubbles may be present in a cubic yard of air-entrained concrete. Figure 7-1 shows a greatly enlarged photograph of hardened air-entrained concrete as it appears through a microscope.

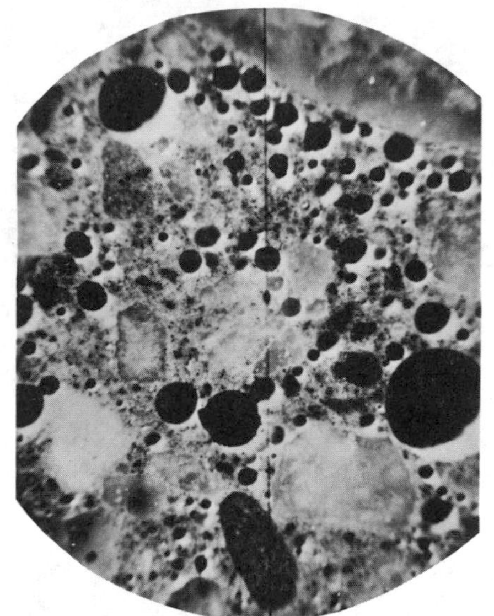

Fig. 7-1 Polished section of air-entrained concrete as seen through a microscope.

EFFECT OF ENTRAINED AIR ON PROPERTIES OF FRESH CONCRETE

The mixing of a cubic yard of air-entrained concrete requires less water than non-air-entrained concrete having the same consistency and maximum aggregate size. Entrained air improves the workability of concrete by increasing the volume of the paste. Lean mixes, which may be harsh and difficult to work, show greatly improved workability with the presence of entrained air. The workability of mixes with angular and poorly graded aggregates is similarly improved. Because of this improved workability, the water and sand content can be reduced significantly.

Fresh concrete containing entrained air is cohesive and looks and feels "fatty." The billions of disconnected air bubbles reduce segregation and bleeding (separation of the water from the paste) of freshly mixed concrete. This reduction in segregation and bleeding results in a more durable concrete. The air bubbles buoy up the aggregates and cement and thus reduce their rate of settling toward the bottom of the placed concrete. Bleeding is retarded because water cannot readily pass through the minute bubbles. Air-entrained concrete can be finished sooner than ordinary concrete. However, the concrete worker should not view the absence of bleeding water as an indication that the concrete is setting and attempt to finish it too soon.

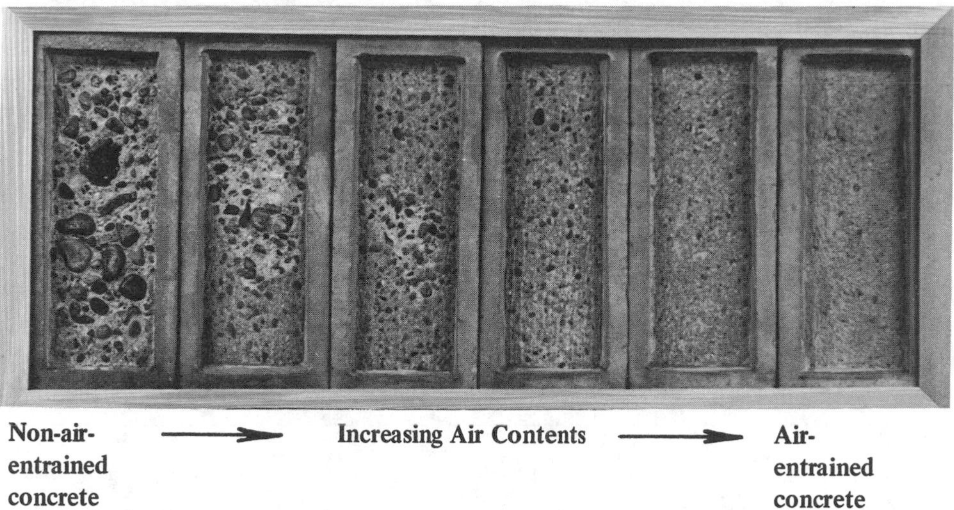

Non-air-entrained concrete ⟶ **Increasing Air Contents** ⟶ **Air-entrained concrete**

Fig. 7-2 Comparison of the results of the freeze-thaw action on test specimens containing varying amounts of air. Note the excellent resistance of the concrete containing adequate entrained air.

EFFECT OF ENTRAINED AIR ON PROPERTIES OF HARDENED CONCRETE

The resistance of concrete to freezing and thawing cycles and to various deicing chemicals is significantly improved by the use of entrained air. This improvement in freeze-thaw resistance is particularly desirable if the concrete is saturated with water. As water freezes, it expands up to 9 percent over its original volume. This expansion produces pressures that can rupture the concrete and cause scaling. Because the entrained air bubbles are round, they never completely fill with water. As a result, when the water freezes, its expansion is relieved by the remaining air space in the entrained bubbles, thus preventing damage to the concrete. The bubbles or air voids continue to serve their purpose during repeated cycles of freezing and thawing. The resistance of air-entrained concrete to the freeze-thaw action is improved several hundred percent over concrete with no entrained air.

Deicing chemicals used for snow and ice removal can cause serious surface scaling. Entrained air prevents salt scaling and is recommended for all concretes that will be in direct contact with deicing chemicals. Indirect application of deicers can occur in various ways, such as drippings from the underside of vehicles.

Concrete mixes that are properly adjusted for the addition of entrained air will have little if any loss of strength when compared to a similar non-air-entrained concrete. It is necessary to adjust the water and sand contents in air-entrained concrete to maintain the same strength. The water content for air-entrained concrete will be 3 to 5 gallons less per cubic yard than for non-air-entrained concrete having the same slump. The sand in air-entrained concrete can be reduced 3 to 6 percent without any loss of workability. If no adjustment is made in the sand content, lower strengths result because of the increased yield.

The abrasion resistance of air-entrained concrete is about the same as that of non-air-entrained concrete of the same compressive strength. Compressive strength is one way to estimate the abrasion resistance of concrete. The abrasion resistance usually increases as the compressive strength increases.

The watertightness of air-entrained concrete is superior to that of non-air-entrained concrete. For a given cement factor, air-entrained concretes generally have lower water-cement ratios and thus are more watertight. Air-entrained concrete is used where watertightness is desired.

AIR-ENTRAINING MATERIALS

The entrainment of air in concrete can be accomplished by adding an air-entraining admixture at the time of mixing, by using an air-entraining cement, or by a combination of both methods.

Many cement manufacturers market portland cement containing air-entraining agents that are interground with the cement during manufacture. Such cements are identified on the bag.

Variations in air content can be expected with variations in aggregate type and size, proportions and gradation, mixing time, temperature, and slump. If air-entraining admixtures are used, the volume of entrained air can be adjusted more readily to meet job conditions by changing the amount of the admixture.

Each of the methods described has certain advantages. The use of air-entraining cement avoids the problem of adding an extra ingredient at the mixer. On jobs where careful control is not possible, air-entraining cements are especially useful since they eliminate the possibility of human or mechanical error when adding an admixture during batching. Regardless of which method is used, adequate control is required to ensure the proper air content at all times.

Recommended Amounts of Entrained Air

Mixes should be designed to contain more air than desired when the concrete is placed. A good rule of thumb is to use 6 percent plus or minus 1 percent.

The aggregate type, gradation, size, and mix proportions affect the percentage of air required. There is little change in air content when the maximum size of aggregate is increased above 1 1/2 inches. For aggregate sizes smaller than 1 1/2 inches, the recommended air content increases as the aggregate size decreases because of the sharp increase in unit mortar content.

Percent Air	Maximum Size Aggregate
5 ± 1	1 1/2 in., 2 in., or 2 1/2 in.
6 ± 1	3/4 in. or 1 in.
7 1/2 ± 1	3/8 in. or 1/2 in.

FACTORS AFFECTING AIR CONTENT

A number of other factors affect the percentage of entrained air in addition to aggregate gradation and mix proportions.

One factor is the fine aggregate content. An increase in the amount of fine aggregate results in a need for more entrained air.

Slump and vibration also affect the amount of air retained in the finished concrete.

The desired air content can be maintained as long as the slump does not exceed 6 inches. Any further increases in slump cause the air content to decrease rapidly.

The duration of vibration also affects air content. A normal amount of vibration does not materially affect the amount of entrained air. However, prolonged vibration of concrete should be avoided. For most concrete, the desired consolidation can be obtained with 5 to 15 seconds vibration. If the vibration is applied properly, little of the intentionally entrained air is lost.

Another factor affecting air content is the concrete temperature. Less air is entrained as the temperature of the concrete increases. The effect of temperature is especially important when concrete is placed in hot weather. A decrease in the air content can be offset, if necessary, by increasing the quantity of air-entraining admixture.

The mixing action also affects the amount of entrained air, especially for ready-mixed concrete. The amount of entrained air varies with the type and condition of the mixer, the amount of concrete being mixed, and the rate of mixing.

MEASUREMENT OF AIR CONTENT

The tests of fresh concrete for air content should be conducted during the actual construction to achieve routine control. Because of the effects of mixing and vibration, samples for air content should be taken from concrete that is placed and consolidated. Unit 9 covers the testing methods standardized by ASTM.

OTHER SPECIAL CONCRETES

Polymer-modified Concrete

A process for producing very high-strength concrete was developed in 1967. This process consists of casting and curing a concrete member, drying it under heat and vacuum, impregnating it with one of a group of organic chemicals called *monomers,* and then causing these monomers to combine into very large molecules called *polymers.* The process of combining monomers is called *polymerization* and is accomplished either by the action of heat plus a catalyst or by radiation.

The increased strength resulting from this process is very large. A concrete which normally has a compressive strength of 3,500 psi can be increased to 17,000 psi if it is manufactured using this process. Initially stronger concretes can reach even higher strengths, with an upper limit of about 24,000 psi. Other properties such as drying shrinkage, tensile strength, resistance to chemical attack, and creep under sustained load are also greatly improved.

To impregnate the concrete completely requires approximately 10 percent monomer by volume of the concrete. The monomer used can be one of many organic chemicals capable of polymerization, including methyl methacrylate, urethane, or styrene. Although polymer impregnation considerably increases the cost of a concrete unit, the improved physical characteristics may outweigh the cost in many different applications.

Latex and epoxies are commonly used polymers. Latex may improve ductility, durability, adhesive properties, resistance to chloride, and other qualities. Bridge overlays and patching commonly use latex-modified concrete. For further information on this subject refer to American Concrete Institute (ACI 548) *Guide for the Use of Polymers in Concrete.*

Wire and Fiber Reinforcement

Fiber reinforced concrete is made from normal portland cement concrete and short fibers. These fibers must be ductile, strong in tension, and capable of bonding to the cement paste. Materials such as glass, nylon, polyethylene, and steel have been used successfully.

Glass fibers are effective because of their high tensile strength. The main problem with most glasses is that they react with the alkalies present in the cement and suffer a reduction in strength. This problem can be overcome by the use of organic coatings of low-alkali cements. The addition of 10 percent by volume of glass fibers can increase the flexural strength of portland cement mortars by a factor of 2. At the same time, the impact resistance of the portland cement mortar is increased by a factor of 10.

Organic fibers such as nylon and polyethylene possess high tensile strength and are not harmed by alkalies. However, these materials have low moduli of elasticity and are not effective in increasing the flexural strength of portland cement mortars and concretes. Nylon and polyethylene do have the ability to increase the impact resistance of mortars and concretes by 10 to 25 times and thus may be useful in blast-resistant structures.

Steel fibers have a high modulus of elasticity, about 10 times that of concrete, and are very strong in tension. The steel fibers bond with the concrete mix reasonably well. The fibers have a high elongation before fracture and are easier to incorporate into a concrete mix. The addition of four percent by volume of steel fibers can double the flexural strength and increase the compressive strength of concrete by 50 percent.

The reinforcing fibers typically are 10 to 20 mils in diameter and range in length from 1/2 to 2 inches. High percentages of fiber tend to produce balls of fiber in a concrete mix. Longer fiber lengths increase this tendency. Workability is always reduced when fibers are added and costs are increased. However, fiber-reinforced concretes have many uses in the concrete industry because of their improved flexural and tensile strengths, higher impact resistance, and crack-arresting characteristics.

STUDY/DISCUSSION QUESTIONS

1. What is the principal reason for using intentionally entrained air in concrete?
2. Explain the effect of entrained air on the properties of fresh concrete.
3. How and why does air entrainment affect workability?
4. How does air entrainment affect the freeze-thaw resistance of concrete? How is this accomplished?
5. Under what climate conditions is air-entrained concrete especially desirable?
6. What are the two methods used to obtain air entrainment in concrete?
7. What is the rule of thumb for determining the percentage of air entrainment required in concrete?
8. Use examples to describe fiber reinforcement.

UNIT 8

SELECTION AND DESIGN OF CONCRETE MIXTURES

The design of concrete mixtures involves the determination of the most economical and practical combination of concrete ingredients to achieve a concrete that is workable in its plastic state and will develop the required qualities when hardened. A properly designed concrete mix achieves the following objectives:

- Required qualities of hardened concrete.
- Workability of fresh concrete.
- Economy.

If acceptable materials are used, the water-cement ratio, the amount of entrained air, and the curing affect the qualities of hardened concrete. These qualities include resistance to freezing and thawing, watertightness, wear resistance, and strength.

The method and length of curing should be determined prior to mixing the concrete so that the proper water-cement ratio can be selected. The air content desired in the concrete depends upon the amount of mortar in the mix; the amount of mortar, in turn, usually depends upon the maximum size of the coarse aggregate.

The workability of fresh concrete is the property that determines the amount of work required to consolidate the concrete completely. Although workability is difficult to measure, it can be readily judged by experienced technicians. Thus, the design of a concrete mix is an art as well as a science.

To achieve economy, the goal of concrete mix design is to minimize the amount of cement required without sacrificing concrete quality. Since quality primarily depends upon the water-cement ratio, the water requirement should be minimized to reduce the cement requirement. Both the water and cement requirements can be minimized by using the stiffest mixture practical, the largest aggregate size practical, and the optimum ratio of fine to coarse aggregates. The relative costs of fine and coarse aggregates should also be considered in determining the most economical mix proportions.

SELECTING MIX CHARACTERISTICS

Before a concrete mixture can be designed, it is necessary to know the size, shape, and required concrete strength of the structure and its exposure conditions. Since most of the desired properties of hardened concrete depend upon the quality of the cement paste, the first step in the design of a concrete mix is the selection of the appropriate water-cement ratio.

Water-Cement Ratio

Table 8-1 can be used as a guide in selecting the water-cement ratio for various exposure conditions. Note that the quantities shown are the *maximum* permissible water-cement ratios.

Under certain conditions, as indicated in Table 8-1, the water-cement ratio should be selected on the basis of concrete strength. In such cases, tests should be made with the

Type of structure	Structure wet continuously or frequently and exposed to freezing and thawing†	Structure exposed to seawater or sulfates
Thin sections (railings, curbs, sills, ledges, ornamental work) and sections with less than 1 inch cover over steel	0.45	0.40‡
All other structures	0.50	0.45‡

† Concrete should also be air-entrained.

‡ If sulfate-resisting cement (Type II or Type V of ASTM C150) is used, the permissible water-cement ratio may be increased by 0.05.

Table 8-1 Maximum permissible water-cement ratios for concrete in severe exposures.*

*Adapted from ACI 318 *Building Code Requirements for Reinforced Concrete.*

materials to be used on the job to determine the relationship between the water-cement ratio and the strength. If data for this relationship cannot be obtained because of time limitations, the water-cement ratio may be estimated from the graphs shown in figure 8-1. A majority of the results of strength tests made by many laboratories using a variety of materials falls within the band curves in this figure.

If the concrete mix design is based on the flexural strength rather than the compressive strength, tests are made to determine the relationship between the water-cement ratio and the flexural strength. An approximate relationship between the flexural and compressive strengths is:

$$f'_c = \left(\frac{R}{K}\right)^2$$

where f'_c = compressive strength, in psi

R = flexural strength (modulus of rupture), in psi

K = a constant, usually between 8 and 10.

If both the exposure conditions and the strength must be considered, the lower of the two indicated water-cement ratios should be used.

Maximum Aggregate Size

The maximum size of coarse aggregate that can be used depends on the size and shape of the concrete members and the amount and distribution of the reinforcing steel. In most cases, the maximum aggregate size should not exceed one-fifth the minimum dimension of the member. Nor should the aggregate size exceed three-fourths the clear space between the reinforcing bars or between the reinforcement and the forms. For unreinforced slabs on ground, the maximum aggregate size should not exceed one-third the slab thickness. Smaller aggregate sizes may be used when availability is a factor and economy must be considered.

The amount of mixing water required to produce a cubic yard of concrete of a given slump (a measure of fluidity) depends upon the maximum size of the aggregate. The smaller the maximum size of the aggregate, the greater is the amount of water required. Therefore,

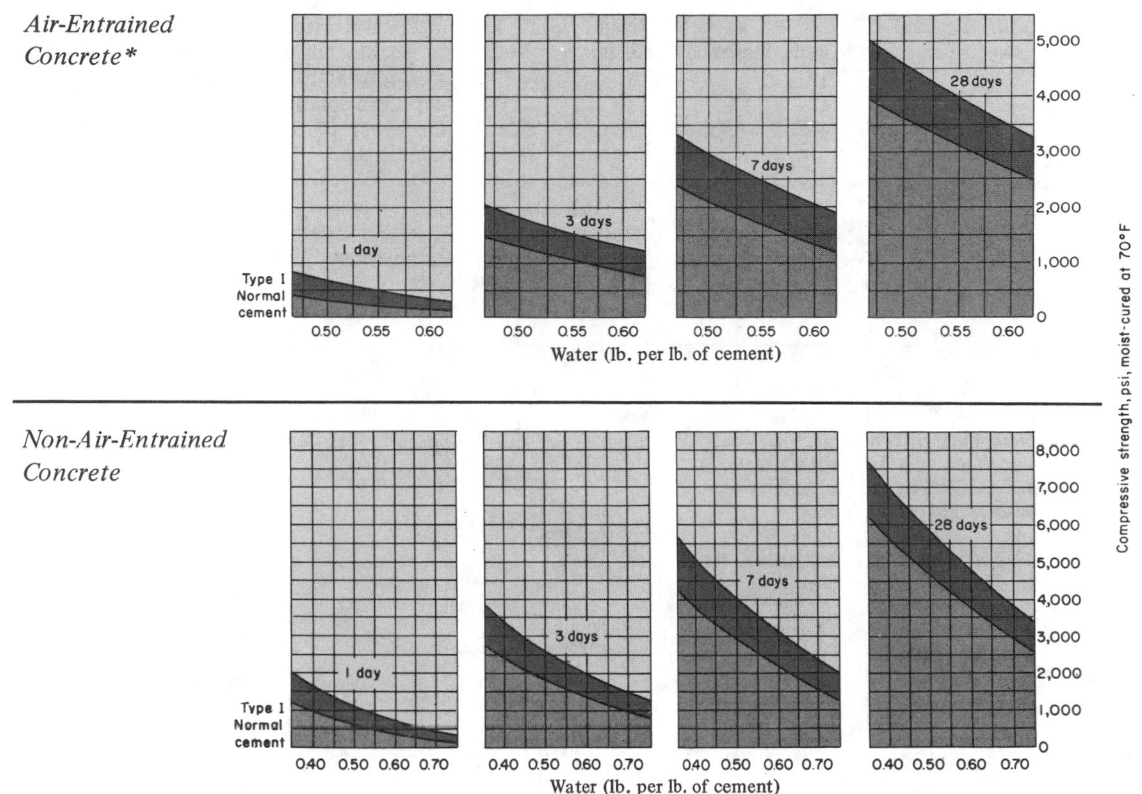

*Concrete with air content within recommended limits and maximum aggregate size of 2 inches or less

Fig. 8-1 Relationship between the water-cement ratio and compressive strength for portland cements at different ages. These relationships are approximate and should be used only as a guide in the absence of data on actual job materials.

it is recommended that the largest practical maximum size of coarse aggregate be used to minimize the water requirement. As a result, the cement content is kept to a minimum.

The maximum size of coarse aggregate used to produce concrete of maximum strength for a given cement content depends upon the aggregate source and the aggregate shape and grading. For many aggregates, this optimum maximum size is about 3/4 inch. If strength is the primary consideration in the design of a concrete mix, it is recommended that the optimum maximum size of aggregate for strength be used.

Air Content

Entrained air should be used in all concrete exposed to freezing and thawing cycles. Entrained air may be used for mild exposure conditions to improve the workability of the concrete. The recommended total air contents for various air-entrained concretes are shown in Table 8-2. Note that the amount of air required to provide adequate freeze-thaw resistance depends upon the maximum size of the aggregate. Entrained air is provided in the mortar fraction of the concrete. In properly designed mixes, the mortar content decreases as the maximum aggregate size increases.

	Water, lb. per cu. yd. of concrete for indicated maximum sizes of aggregate						
Slump, in.	3/8	1/2	3/4	1	1 1/2	2	3
	Non-air-entrained concrete						
1 to 2	350	335	315	300	275	260	220
3 to 4	385	365	340	325	300	285	245
6 to 7	410	385	360	340	315	300	270
Air content, percent	3	2.5	2	1.5	1	0.5	0.3
	Air-entrained concrete						
1 to 2	305	295	280	270	250	240	205
3 to 4	340	325	305	295	275	265	225
6 to 7	365	345	325	310	290	280	260
Air content, percent (Severe Exposure)	8	7	6	6	5.5	5	4.5

Table 8-2 Approximate mixing water and air content requirements for different slumps and maximum sizes of aggregates.*

*Adapted from ACI 318 *Building Code Requirements for Reinforced Concrete.*

These quantities of mixing water are for use in computing cement contents for trial batches. They are the maximum amounts for reasonably well-shaped angular coarse aggregates graded within the limits of accepted specifications.

Slump

The slump test is used as a measure of the fluidity of the concrete. This test should not be used to compare mixes of wholly different proportions or mixes of different kinds or sizes of aggregates. Under conditions of uniform operation, changes in slump indicate changes in materials, mix proportions, or water content. Slumps within the limits given in Table 8-3, are suggested to avoid mixes that are too stiff or too fluid.

TRIAL MIX METHOD

The most direct method for determining the optimum mix proportions is through the use of trial mixes. Such mixes may be relatively small batches precisely proportioned or job-size batches made during the course of normal concrete production. The use of a combination of both methods often achieves the most satisfactory results. For either method, the water-cement ratio, the maximum size of aggregate, the air content, and the range of slump are the factors to be selected first. By maintaining each of these factors constant

Types of Construction	Slump, in.	
	Maximum*	Minimum
Reinforced foundation walls and footings	3	1
Plain footings, caissons, and substructure walls	3	1
Beams and reinforced walls	4	1
Building columns	4	1
Pavements and slabs	3	1
Heavy mass concrete	2	1

*May be increased 1 inch for methods of consolidation other than vibration.

Table 8-3 Recommended slumps for various types of construction.

*Adapted from ACI 211.1 *Standard Practice for Selecting Proportions for Normal, Heavyweight and Mass Concrete.*

while the amounts of fine and coarse aggregates are varied, a series of mixes can be made. The proportions of the ingredients are computed for each mix and the results are compared. Based on workability and economic considerations, the mix proportions are selected.

Laboratory Trial Mixes

To illustrate the small lot trial mix procedure, assume that the mix proportions are to be determined for the following application.

Conditions. Concrete is required for a retaining wall that will be exposed to fresh water in a severe climate. A design compressive strength of 3,000 psi at 28 days is specified. The minimum thickness of the wall is 8 inches and 2 inches of concrete cover over the reinforcement is required.

Water-cement ratio. For the exposure conditions given, Table 8-1 indicates that air-entrained concrete with a maximum water-cement ratio of 0.50 per bag should be used. Figure 8-1 indicates that the water-cement ratio required for strength should range between 0.51 and 0.60. Refer to the graph for the 28-day strength of Type I cement. The ratio is used assuming that the mix is designed for 3,450 psi (3,000 + 15 percent*) to allow for job variations. These water-cement ratios are greater than the ratio permitted for the exposure conditions. Because of the exposure requirements, not more than 0.50 should be used. It should be noted that this water-cement ratio will produce strengths higher than needed for structural purposes.

Aggregate size. From the information given, the maximum size of the aggregates used should not exceed 1 1/2 inches. Assume that satisfactory 1 1/2-inch maximum size aggregate is economically available. If this size is not available, then the next smaller available size is used.

Air content. Because of the severe exposure conditions, the concrete should contain entrained air. Table 8-2 indicates that the recommended air content for concrete using 1 1/2-inch aggregate is 4.5 percent.

The entrained air is provided using either an air-entraining portland cement or an air-entraining admixture. Unless an air-entraining cement is used, sufficient air-entraining admixture must be added to a portion of the mixing water to insure the desired air content. In general, the amount of admixture recommended by the manufacturer will produce the desired air content.

Slump. A typical slump range for a reinforced foundation wall is 1 to 3 inches, assuming that the concrete will be consolidated by vibration, Table 8-3.

Note: All of the trial mix data are entered in the appropriate blanks on the Trial Mix Data Sheet, figure 8-2. Several of these blank sheets are included in the Appendix for student use.

Representative samples of the aggregates, cement, water, and air-entraining admixture must be used. To simplify the calculations and to eliminate sources of error caused by variations in the moisture content of the aggregate, the aggregates should be prewetted and allowed to dry to a saturated, surface-dry (SSD) condition. The aggregates should be placed in covered containers to maintain this condition until use.

The size of the trial batch depends upon the equipment available and the number and size of test specimens to be made. If the batches are to be mixed by hand and if no compression test specimens are required, a 10-pound batch may be adequate; however, larger batches will produce more accurate data. Machine mixing is recommended since it more nearly represents job conditions. The concrete must be mixed by machine if it is to contain entrained air.

Batch weights. For convenience, assume that the batch containing 20 lb. of cement is to be made. The quantity of mixing water required is 20 x 0.50 or 10.0 lb. These quantities of cement and water are carefully weighed. Representative samples of fine and coarse aggregates in suitable containers are weighed. These weights are indicated as the initial weights in column 2 of the data sheet, figure 8-2.

All of the measured quantities of cement, water, and air-entraining admixture are to be used. The fine and coarse aggregates are added to produce a workable concrete mixture having a slump adequate for placement in a retaining wall. An experienced technician can determine if the slump is adequate by noting if the ingredients tend to segregate. The technician can also observe the workability of the concrete. By adding fine and coarse aggregates in small increments, the technician controls the slump.

The use of too much coarse aggregate will produce a harsh unworkable mix. Too much fine aggregate will result in an uneconomical mix with high water and cement requirements and greater potential volume changes. A rule of thumb is to use as much coarse aggregate as possible with sufficient fine aggregate to insure adequate workability. Mechanical mixing, simulating the type to be used on the project, should be used to obtain the most reliable results.

Workability. When the slump of the concrete appears to be within the recommended range, tests for slump, air content, and unit weight are made. The results of these tests are noted on the data sheet as well as a description of the workability. (Refer to Table 8-4.) The weights of the fine and coarse aggregates that are not used are recorded on the data sheet in column 3, and the weights of materials used are noted in column 4 (column 2 minus column 3),

Data and Calculations for Trial Batch
(Saturated, surface-dry aggregates)

BATCH SIZE: 100 ____ 50 ____ 20 ✓ 10 ____ lb. cement

Note: Complete columns 1 through 5, fill in items below, then complete columns 6 and 7

(1) Material	(2) Initial wt. (lb.)	(3) Final wt. (lb.)	(4) Wt. used (col. 2 minus col. 3)	(5) Wt. of batch for 100 lb. cement (col. 4 x 100 ÷ batch size)	(6) Wt./cu. yd. (col. 5 x cement content)	(7) Remarks
Cement	20.0	0	20.0	100	543	*Cement factor 5.43 cwt./cu. yd.
Water	10.0	0	10.0	50	272	
Fine Aggregate	66.2	28.2	38.0	190	1032	**Fine aggregate 33.6% of total aggregate
Coarse Aggregate	89.8	14.0	75.8	376	2040	
A/E Admixture	0.30 oz.	Total (T) = 716				

Measured Slump: 3 in. Measured Air Content: 5.4 percent

Appearance: Sandy ____ Good ✓ Rocky ____

Workability: Good ✓ Fair ____ Poor ____

Wt. of Container + Concrete = 42.6 lb.

Wt. of Container = 6.6 lb.

Wt. of Concrete (A) = 36.0 lb.

Vol. of Container (B) = 0.25 cu. ft.

UNIT WT. of CONCRETE (w) = $\frac{A}{B} = \frac{36.0}{0.25} =$ 144 lb./cu. ft.

YIELD = $\frac{\text{Total Wt. of material for 100 lb. cement}}{\text{Unit wt. of concrete}} = \frac{T}{W} = \frac{716}{144} =$ 4.97 cu. ft./cwt.

*CEMENT FACTOR = $\frac{\text{27 cu. ft.}}{\text{yield}} = \frac{27}{4.97} =$ 5.43 cwt./cu. yd.

**Fine Aggregate % of Total Aggregate = $\frac{\text{Wt. of fine aggregate}}{\text{Total wt. of aggregates}} = \frac{1032}{3072} =$ 33.6 %

Fig. 8-2 Data Sheet for laboratory trial mixes.

figure 8-2. Since the batch is based on a 20-lb. mix, the quantities for a 100-lb. mix are equal to the weights used multiplied by 5.

If the slump is greater than that required, small quantities of fine or coarse aggregate (or both) are added. If the slump is less than that required, the appropriate ratio of water and cement is added to produce the desired slump. In this situation, it is important that the quantities be measured accurately and then recorded on the data sheet.

Finding the yield and cement factor. The yield is the volume of concrete produced from a mixture containing 100 lb. of cement. The yield is equal to the weight of a 1-cwt. batch divided by the unit weight of the concrete. The number of hundredweights of cement required per cubic yard of concrete (cement factor) is computed by dividing 27 (cubic feet per cubic yard) by the yield in cubic feet. The percentage of fine aggregate of the total aggregate is also calculated. For the trial laboratory batch in this example, the cement factor is 5.43 cwt./cu. yd., the unit weight of the concrete is 144 pcf, and the fine aggregate is 33.6 percent of the total aggregate by weight.

To determine the most economical proportions, additional trial batches should be made by varying the percentage of the fine aggregate. In each batch, the water-cement ratio, aggregate gradations, air content, and slump are maintained at approximately the same values. The results of four trial batches are summarized in Table 8-4.

Quantities. For the trial mixes in this example, the percentage of fine aggregate is plotted against the cement content in figure 8-3. Note that for the combination of materials used, the minimum cement factor, 5.43 cwt./cu. yd., occurs at a fine aggregate content of about 32 percent of the total aggregate. Since the water-cement ratio is 0.50 and the unit weight of the concrete for an air content of 5 percent is about 144 pcf, the quantities of materials required per cubic yard can be computed:

Cement	5.25 cwt./cu.yd.	= 525 lb.
Water	5.25 x 0.50	= 273 lb.
Cement + Water		= 798 lb.
Wt. concrete per cu. yd.		= 144 x 27 = 3,888 lb.
Wt. aggregates		= 3,888 – 798 = 3,090 lb.
Wt. fine aggregate		= (32/100) x 3,090 = 990 lb.
Wt. coarse aggregate		= 3,090 – 900 = 2,100 lb.

Batch No.	Slump, in.	Air content, percent	Unit wt., pcf	Cement factor cwt./cu. yd.	Fine aggregate percent of total aggregate	Workability
1	3	5.4	144	5.43	33.6	Excellent
2	2 3/4	4.9	144	5.57	27.4	Harsh
3	2 1/2	5.1	144	5.59	36.1	Excellent
4	3 1/4	4.7	145	5.43	30.5	Good

Table 8-4 Example of results of laboratory trial mixes.

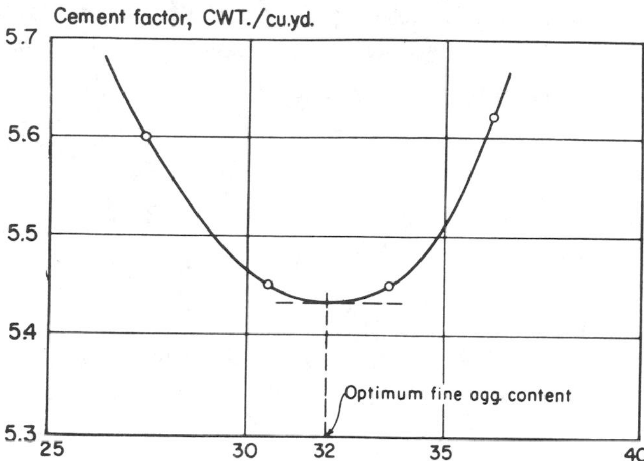

Fig. 8-3 Example of relationship between percentage of fine aggregate and cement content for a given water-cement ratio and slump.

Unless care is taken to control the slump and air content and unless weights are accurately determined, it may be difficult to obtain the data required to plot a curve such as the one shown in figure 8-3. For well-graded, rounded aggregates, the curve may be nearly flat. If there is a wide difference in the costs of fine and coarse aggregates, the optimum percentage of fine aggregate required for an economical mix may be different than the percentage at which the minimum cement factor occurs. For example, if coarse aggregate costs more than fine aggregate, it may be more economical to use more fine aggregate, with an appropriate increase in the cement content.

Job-Size Trial Batches

On most jobs, trial batches may be full size. These batches are often used in the foundations during the early phases of a construction project. The first trial mix may be selected on the basis of experience or from established relationships such as those given in Tables 8-5 and 8-6. These tables are based on experience and data from several sources. Tables 8-5 and 8-6 indicate the amount of water required per cubic yard of concrete, the cement content, and the proportion of fine aggregate required for good workability.

Values in these tables are based on concrete having a slump of 3 to 4 inches. The concrete contains well-graded aggregates having a specific gravity of 2.65. For other conditions, it is necessary to recalculate the quantities following the guidelines given in the footnotes. When making a trial batch, enough water should be added to produce the desired slump. This amount of water, however, may be more or less than the estimated amount. The slump, air content, and the unit weight of the fresh concrete then can be measured.

To illustrate the method of making a job-size trial batch, the mix proportions for the following application are to be determined.

Conditions. For this concrete mix, the water-cement ratio is 0.55, the maximum aggregate size is 1 inch, the air content is 6 ± 1 percent, and the slump is 2 to 4 inches. The fine aggregate has a fineness modulus of 2.85 and a moisture content of 4.3 percent. The coarse aggregate has a moisture content of 1 percent.

Weight. For the above conditions, Table 8-5 provides the following values per cubic yard of concrete:

Cement	5.1 cwt.	=	510 lb.
Water		=	280 lb.
Fine aggregate		=	1,165 lb.
Coarse aggregate		=	1,835 lb.
	Total	=	3,790 lb.

The free moisture (above the saturated, surface-dry condition) in the fine aggregate amounts to 4.3 percent of 1,165 lb., or 50 lb. The free moisture in the coarse aggregate is approximately 18 lb. The corrected weights per cubic yard of concrete are:

Cement (Type IA)	510 + 0	=	510 lb.
Water	280 – 50 – 18	=	212 lb.
Fine aggregate	1,165 + 50	=	1,215 lb.
Coarse aggregate	1,835 + 18	=	1,853 lb.
	Total	=	3,790 lb.

When the above material quantities are mixed, the consistency is such that the slump is less than 1 inch. Additional water (25 lb.) is added to bring the slump to 3 inches. The air content is measured to be 5.3 percent and the unit weight is measured to be 144 pcf. The workability of the mix is good. The total amount of water used is 305 lb. (280 + 25); the total weight of concrete produced is 3,815 lb. (3,790 + 25).

Volume. Since the unit weight of the concrete is 144 pcf, the volume of the concrete in this batch is 26.3 cu. ft. (3,790 ÷ 144 pcf).

The principal reason for the discrepancy between the actual and the theoretical yield is that the specific gravity values of the aggregates may be different from those assumed in Tables 8-5 and 8-6. As a result, these quantities must be adjusted to produce 27 cu. ft. of concrete and maintain the water-cement ratio at 0.55. Unless the corrections are very large, the following simplified assumptions can be made. The unit weight of the concrete is assumed to remain essentially constant and the amount of water required (per cubic yard of concrete) also is assumed to remain constant.

Adjustment. For example, the adjusted water requirement for this application is:

$$\frac{27}{26.3} \times 305 = 314 \text{ lb.}$$

The adjusted cement requirement is: $\frac{314}{0.55} = 571$ lb. The weight of the materials per cubic yard of concrete must total 144 x 27 = 3,888 lb. The total weight of the aggregates must equal 3,888 – (314 + 571) = 3,003 lb. For this ratio of fine to coarse aggregates, the saturated surface-dry weights are 1,170 lb. and 1,833 lb., respectively.

To determine the optimum proportions to achieve the economic use of the materials, additional trial batches should be made by varying the ratio of fine and coarse aggregates, as covered in the preceding section on Laboratory Trial Mixes.

See Table 8–7 for admixtures for concrete.

					With fine sand fineness modulus = 2.50**			With coarse sand fineness modulus = 2.90**		
Water/ cement ratio, by wt.	**Maxi-mum size of aggre-gate, in.**	**Air content, percent**	**Water, lb. per cu. yd. of concrete**	**Cement, cwt. per cu. yd. of concrete**	**Fine ag-gregate, percent of total aggregate**	**Fine ag-gregate, lb. per cu. yd. of concrete**	**Coarse ag-gregate, lb. per cu. yd. of concrete**	**Fine ag-gregate, percent of total aggregate**	**Fine ag-gregate, lb. per cu. yd. of concrete**	**Coarse ag-gregate, lb. per cu. yd. of concrete**
0.4	3/8	7.5	340	8.5	53	1320	1190	57	1430	1080
0.4	1/2	7.5	320	8.0	43	1120	1460	48	1230	1350
0.4	3/4	6	300	7.5	36	1010	1760	40	1120	1650
0.4	1	6	280	7.0	32	910	1940	36	1020	1820
0.4	1 1/2	5	270	6.8	28	820	2150	32	950	2030
0.45	3/8	7.5	340	7.6	54	1390	1190	58	1500	1080
0.45	1/2	7.5	320	7.1	45	1190	1460	49	1300	1350
0.45	3/4	6	300	6.7	38	1080	1760	42	1190	1650
0.45	1	6	280	6.2	33	970	1940	37	1090	1820
0.45	1 1/2	5	270	6.0	29	880	2150	33	1000	2030
0.5	3/8	7.5	340	6.8	55	1440	1190	59	1550	1080
0.5	1/2	7.5	320	6.4	46	1250	1460	50	1360	1350
0.5	3/4	6	300	6.0	39	1130	1760	43	1240	1650
0.5	1	6	280	5.6	34	1020	1940	38	1140	1820
0.5	1 1/2	5	270	5.4	30	930	2150	34	1050	2030
0.55	3/8	7.5	340	6.2	56	1490	1190	60	1600	1080
0.55	1/2	7.5	320	5.8	47	1300	1460	51	1410	1350
0.55	3/4	6	300	5.5	40	1170	1760	44	1280	1650
0.55	1	6	280	5.1	35	1060	1940	39	1180	1820
0.55	1 1/2	5	270	4.9	31	960	2150	35	1080	2030
0.6	3/8	7.5	340	5.7	56	1540	1190	60	1650	1080
0.6	1/2	7.5	320	5.3	48	1340	1460	52	1450	1350
0.6	3/4	6	300	5.0	41	1210	1760	44	1310	1650
0.6	1	6	280	4.7	36	1090	1940	40	1210	1820
0.6	1 1/2	5	270	4.5	32	990	2150	35	1110	2030
0.65	3/8	7.5	340	5.2	57	1570	1190	61	1680	1080
0.65	1/2	7.5	320	4.9	48	1370	1460	52	1480	1350
0.65	3/4	6	300	4.6	41	1240	1760	45	1350	1650
0.65	1	6	280	4.3	37	1120	1940	41	1240	1820
0.65	1 1/2	5	270	4.2	32	1020	2150	36	1140	2030
0.7	3/8	7.5	340	4.9	57	1600	1190	61	1710	1080
0.7	1/2	7.5	320	4.6	49	1400	1460	53	1510	1350
0.7	3/4	6	300	4.3	42	1270	1760	46	1380	1650
0.7	1	6	280	4.0	37	1150	1940	41	1260	1820
0.7	1 1/2	5	270	3.9	33	1050	2150	37	1170	2030
0.75	3/8	7.5	340	4.5	58	1620	1190	62	1740	1080
0.75	1/2	7.5	320	4.3	49	1420	1460	53	1530	1350
0.75	3/4	6	300	4.0	42	1280	1760	46	1400	1650
0.75	1	6	280	3.7	38	1170	1940	42	1290	1820
0.75	1 1/2	5	270	3.6	33	1070	2150	37	1190	2030

*Increase or decrease water per cubic yard by 3 percent for each increase or decrease of 1 inch in slump. For manufactured fine aggregate, increase percentage of fine aggregate by 3 and water by 15 lb. per cubic yard of concrete. For less workable concrete, as in pavements, decrease the percentage of fine aggregate by 3 and water by 10 lb. per cubic yard of concrete.

**For definition of fineness modulus, see *Aggregates for Concrete,* available only in the United States and Canada on request from the Portland Cement Association.

Table 8-5 Suggested trial mixes for air-entrained concrete of medium consistency (3- to 4-inch slump*).

					With fine sand fineness modulus = 2.50**			With coarse sand fineness modulus = 2.90**		
Water/cement ratio, by wt.	Maximum size of aggregate, in.	Air content (entrapped air), percent	Water, lb. per cu. yd. of concrete	Cement, cwt. per cu. yd. of concrete	Fine aggregate, percent of total aggregate	Fine aggregate, lb. per cu. yd. of concrete	Coarse aggregate, lb. per cu. yd. of concrete	Fine aggregate, percent of total aggregate	Fine aggregate, lb. per cu. yd. of concrete	Coarse aggregate, lb. per cu. yd. of concrete
0.4	3/8	3	380	9.5	53	1320	1190	57	1430	1080
0.4	1/2	2.5	360	9.0	44	1150	1460	48	1260	1350
0.4	3/4	2	340	8.3	36	1000	1760	40	1110	1650
0.4	1	1.5	320	8.0	32	910	1940	36	1030	1820
0.4	1 1/2	1	300	7.5	28	830	2150	32	970	2030
0.45	3/8	3	380	8.5	54	1400	1190	58	1510	1080
0.45	1/2	2.5	360	8.0	46	1230	1460	50	1340	1350
0.45	3/4	2	340	7.6	38	1070	1760	42	1180	1650
0.45	1	1.5	320	7.1	34	990	1940	38	1100	1820
0.45	1 1/2	1	300	6.7	30	910	2150	34	1030	2030
0.5	3/8	3	380	7.6	55	1460	1190	59	1570	1080
0.5	1/2	2.5	360	7.2	47	1290	1460	51	1400	1350
0.5	3/4	2	340	6.8	39	1120	1760	43	1230	1650
0.5	1	1.5	320	6.4	35	1040	1940	39	1160	1820
0.5	1 1/2	1	300	6.0	31	960	2150	35	1080	2030
0.55	3/8	3	380	6.9	56	1520	1190	60	1630	1080
0.55	1/2	2.5	360	6.5	48	1340	1460	52	1450	1350
0.55	3/4	2	340	6.2	40	1170	1760	44	1280	1650
0.55	1	1.5	320	5.8	36	1090	1940	40	1210	1820
0.55	1 1/2	1	300	5.5	32	1000	2150	36	1120	2030
0.6	3/8	3	380	6.3	57	1550	1190	61	1660	1080
0.6	1/2	2.5	360	6.0	49	1380	1460	53	1500	1350
0.6	3/4	2	340	5.7	41	1220	1760	45	1330	1650
0.6	1	1.5	320	5.3	37	1130	1940	41	1250	1820
0.6	1 1/2	1	300	5.0	32	1030	2150	36	1160	2030
0.65	3/8	3	380	5.9	58	1610	1190	61	1720	1080
0.65	1/2	2.5	360	5.5	49	1430	1460	53	1540	1350
0.65	3/4	2	340	5.2	42	1250	1760	45	1360	1650
0.65	1	1.5	320	4.9	37	1160	1940	41	1280	1820
0.65	1 1/2	1	300	4.6	33	1060	2150	37	1190	2030
0.7	3/8	3	380	5.4	58	1640	1190	62	1750	1080
0.7	1/2	2.5	360	5.2	50	1460	1460	54	1570	1350
0.7	3/4	2	340	4.9	42	1280	1760	46	1390	1650
0.7	1	1.5	320	4.6	38	1190	1940	42	1310	1820
0.7	1 1/2	1	300	4.3	34	1100	2150	38	1220	2030
0.75	3/8	3	380	5.1	58	1670	1190	62	1780	1080
0.75	1/2	2.5	360	4.8	51	1490	1460	54	1600	1350
0.75	3/4	2	340	4.5	42	1300	1760	46	1410	1650
0.75	1	1.5	320	4.3	38	1210	1940	42	1330	1820
0.75	1 1/2	1	300	4.0	34	1120	2150	38	1240	2030

*Increase or decrease water per cubic yard by 3 percent for each increase or decrease of 1 inch in slump. For manufactured fine aggregate, increase percentage of fine aggregate by 3 and water by 15 lb. per cubic yard of concrete. For less workable concrete, as in pavements, decrease the percentage of fine aggregate by 3 and water by 10 lb. per cubic yard of concrete.

**For definition of fineness modulus, see *Aggregates for Concrete,* available only in the United States and Canada on request from the Portland Cement Association.

Table 8-6 Suggested trial mixes for non-air-entrained concrete of medium consistency (3- to 4-inch slump*).

Type of admixture	Desired effect	Material
Accelerators (ASTM C 494, Type C)	Accelerate setting and early-strength development	Calcium chloride (ASTM D98) Triethanolamine, sodium thiocyanate, calcium formate, calcium nitrite, calcium nitrate
Air-entraining admixtures (ASTM C 260)	Improve durability in environments of freeze-thaw, deicers, sulfate, and alkali reactivity Improve workability	Salts of wood resins (Vinsol resin) Some synthetic detergents Salts of sulfonated lignin Salts of petroleum acids Salts of proteinaceous material Fatty and resinous acids and their salts Alkylbenzene sulfonates Salts of sulfonated hydrocarbons
Coloring agents	Colored concrete	Modified carbon black, iron oxide, phthalocyanine, umber, chromium oxide, titanium oxide, cobalt blue (ASTM C 979)
Corrosion inhibitors	Reduce steel corrosion activity in a chloride environment	Calcium nitrite, sodium nitrite, sodium benzoate, certain phosphates or fluosilicates, fluoaluminates
Dampproofing admixtures	Retard moisture penetration into dry concrete	Soaps of calcium or ammonium stearate or oleate Butyl stearate Petroleum products
Fungicides, germicides, and insecticides	Inhibit or control bacterial and fungal growth	Polyhalogenated phenols Dieldrin emulsions Copper compounds
Pumping aids	Improve pumpability	Organic and synthetic polymers Organic flocculents Organic emulsions of paraffin, coal tar, asphalt, acrylics Bentonite and pyrogenic silicas Natural pozzolans (ASTM C 618, Class N) Fly ash (ASTM C 618, Classes F and C) Hydrated lime (ASTM C 141)
Retarders (ASTM C 494, Type B)	Retard setting time	Lignin Borax Sugars Tartaric acid and salts

Table 8-7 Admixtures for concrete.*

*For further studies and other admixtures refer to the ASTM references as indicated.

(continued)

Type of admixture	Desired effect	Material
Superplasticizers* (ASTM C 1017, Type 1)	Flowing concrete Reduce water-cement ratio	Sulfonated melamine formaldehyde condensates Sulfonated naphthalene formaldehyde condensates Lignosulfonates
Superplasticizer* and retarder (ASTM C 1017, Type 2)	Flowing concrete with retarded set Reduce water	See Superplasticizers and also Water reducers
Workability agents	Improve workability	Air-entraining admixtures Finely divided admixtures, except silica fume Water reducers

Table 8-7 Continued

*For further studies and other admixtures refer to the ASTM references as indicated.

STUDY/DISCUSSION QUESTIONS

1. What is meant by the design of a concrete mixture?
2. What are the three objectives of a properly designed concrete mixture?
3. Explain what economy refers to in mix design.
4. Name the three most direct trial mix methods for determining the optimum mix proportions.
5. Explain what you feel is the best size for a trial batch.
6. How can air be entrained in a trial batch?
7. Excessive coarse aggregate will produce what type of mix?
8. Excessive fine aggregate will produce what type of mix?
9. Explain two methods for selecting trial batches.
10. How can the optimum proportions be determined to achieve economy of materials?

UNIT 9

SAMPLING AND TESTING PLASTIC CONCRETE

Complete coverage of all of the methods for sampling and testing plastic and hardened concrete is too lengthy for inclusion in this unit. However, a general knowledge of sampling and testing plastic or fresh concrete will benefit all concrete workers.

The need for standardization in testing led to the formation of the American Society for Testing and Materials (ASTM). All tests should be made in accordance with the standardized methods of the ASTM.

SAMPLING FRESH CONCRETE

Specification ASTM C172, *Method of Sampling Freshly Mixed Concrete,* covers the procedure for obtaining samples of fresh concrete from stationary and paving mixers. and from truck mixers, agitators, or dump trucks.

The sample must consist of not less than 1 cu. ft. of concrete when it is to be used for strength tests. Smaller samples are permitted for routine air content and slump tests.

The sampling procedures in the following list shall include the use of every precaution to insure that the samples obtained will be representative of the true nature and condition of the concrete sampled.

(a) Stationary mixers, except paving mixers: pass a receptacle through the discharge stream, or about the middle of the batch.

(b) Paving mixers: after the mixer has discharged the concrete on subgrade, collect portions from a sufficient number of points to be representative.

(c) Revolving drum truck mixers or agitators: sample concrete at two or more regularly spaced intervals during discharge of the middle portion of the batch. Sampling is done by repeatedly passing a receptacle through the discharge stream or by diverting the stream completely so that it discharges into a container such as a wheelbarrow.

(d) Open-top truck mixers, agitators, dump trucks, or other types of open-top containers: samples must be taken using procedures (a), (b), or (c) above.

The sample must be transported to the area where the test is to be made. The sample is then remixed with a shovel just enough to insure uniformity. The sample must be protected from sunlight and wind during the period between the gathering of the sample and the test. This period must not exceed 15 minutes.

TESTS FOR CONSISTENCY

ASTM Designation C143, *Test Method for Slump of Portland Cement Concrete,* covers the procedure for determining the slump of concrete both in the laboratory and in the field.

The slump test may be used as a rough measure of the consistency of concrete, that is, how wet or dry the concrete is. This test is not to be considered as a measure of workability or the proper water content. This test should not be used to compare mixes of entirely different proportions or containing different kinds of aggregates. Any change in slump on the job indicates that changes were made in the grading or proportions of the aggregates or in the water content. The mix should be corrected immediately to obtain the proper consistency. These corrections are obtained by adjusting the amounts and proportion of the sand and coarse aggregate used. There should be no change in the total amount of mixing water specified for each bag of cement.

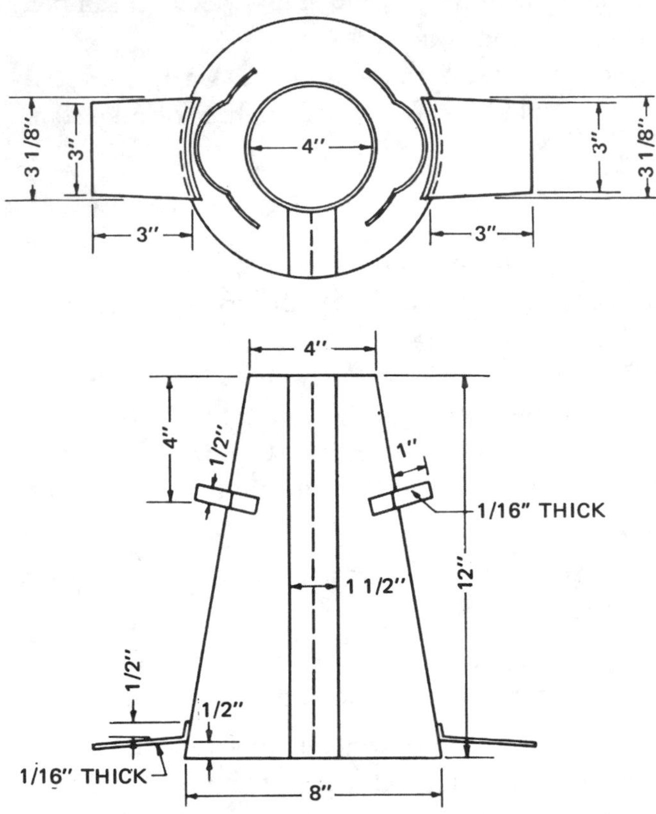

Fig. 9-1 Mold for slump test.

To make the slump test, the test specimen is made in a mold or *slump cone* of 16-gage galvanized metal in the form shown in figure 9-1. The cone is 8 inches in diameter at the base and 4 inches in diameter at the top. The height of the cone is 12 inches. The base and top are open. The mold is provided with foot pieces and handles as shown.

The concrete sample for a slump test should be taken just before the concrete is placed in the forms. The mold is dampened and placed on a flat surface such as a smooth plank or slab of concrete, figure 9-2. The mold is filled with concrete to about one-third its capacity. The concrete is then *rodded* with 25 strokes of a 5/8-inch steel rod about 24 inches long and having a hemispherical tip. The filling is completed in two more layers. Each layer is rodded 25 times and each rod stroke must penetrate into the underlying layer of concrete. After the top layer is rodded, it is struck off using a screeding and rolling motion of the tamping rod. The mold then is carefully removed by raising it vertically in 5 seconds plus or minus 2 seconds. The mold is gently placed beside the concrete specimen and the slump is measured at the displaced original center of the top surface of the specimen as shown in figure 9-3. The test is run within an elapsed time of 2 1/2 minutes. For example, assume that the center of the top of the slumped pile of concrete is measured to be 5 inches below the top of the cone. Thus, the slump of this specimen of concrete is 5 inches.

BALL PENETRATION TEST

A convenient supplementary test to determine the consistency of concrete is covered by ASTM Designation C360, *Test Method for Ball Penetration in Fresh Portland Cement Concrete.* This test is not a substitute for the slump test. Also referred to as the *Kelly*

Fig. 9-2 The slump test shows the consistency of concrete. Here the rodding of the concrete in the cone insures complete filling of the mold.

Fig. 9-3 Slump is measured from a rod laid across the top of the slump cone. The amount of slump shown indicates a medium-wet concrete mixture.

Ball test, this test involves the determination of the depth to which a 6-inch diameter metal hemisphere, weighted to 30 ± 0.1 lb., will sink into the fresh concrete. This test does not require the molding of a specimen. As a result, the concrete can be tested in a wheelbarrow or other container, or even in the forms. The test can be made quickly. It provides a convenient means for the routine checking of the consistency of concrete for control purposes. The results cannot be translated quantitatively into terms of slump and should not be used as a basis for the acceptance or rejection of the concrete.

To avoid the effects of boundary conditions, the ball penetration test must be made on a reasonably large sample. ASTM C360 requires a sample depth of at least 8 inches of concrete and a minimum lateral dimension of about 18 inches. The sample must not be disturbed or vibrated during the test. Care must be taken that the weight is allowed to settle freely but without impact. The reported results should be an average of at least three readings.

Fig. 9-4 Air indicator.

UNIT WEIGHT TEST

The standard test for the unit weight of fresh concrete (ASTM C138) is similar to the test used for the unit weight of aggregates (ASTM C29). The same measuring containers are used with the following exceptions for the concrete test: the 1/2-cu. ft. measure is used when the nominal maximum aggregate size does not exceed 2 inches and the 1-cu. ft. measure is used when the maximum aggregate size is greater than 2 inches. The concrete is placed in three approximately equal layers. Each layer is rodded and the side of the measure is then tapped with the rod 10 to 15 times or until no large bubbles of air appear on the layer. The measure is filled completely and is struck off level with a flat cover plate. Excess concrete is cleaned off and the net weight of the concrete is determined.

TESTING FOR AIR CONTENT

There are a number of methods for determining the air content of fresh concrete. The following three methods have been standardized by the ASTM.

The Pressure Method (ASTM C231)

This method is practical for the field testing of all concretes except those made with highly porous and lightweight aggregates.

The Volumetric Method (ASTM C173)

This method is practical for the field testing of all concretes. However, it is particularly useful for concrete made with lightweight and porous aggregates.

The Gravimetric Method (ASTM C138)

An accurate knowledge of specific gravities and absolute volumes of concrete ingredients is required for this method. It is impractical as a field test method but can be used satisfactorily in the laboratory.

Unit Weight Field Test

A field test for checking yield and possible changes in air content or mix proportions is the unit weight test. The only equipment needed to perform this test is a sturdy container of known volume, preferably 1/4 or 1/2 cu. ft., and an accurate balance. Changes in air content generally result in changes in the unit weight of concrete from one batch to another.

The approximate air content of fresh concrete can be checked with a pocket-size air indicator, figure 9-4. The small container is filled with a representative sample of mortar from the fresh concrete. The tube is then filled with alcohol. After placing the thumb over the open end, the tube is shaken to remove the air from the mortar. The approximate air content is determined from the drop in the level of the alcohol. Since this test can be performed in a few minutes, it is especially useful for estimating the air content of fresh concrete. However, this test serves as an indication only and is not a substitute for the more accurate pressure and volumetric methods.

FOLLOW-UP TESTS

Follow-up tests to determine the effectiveness of the field control methods are made on many jobs. These tests are usually made to find the compressive strength or the flexural strength. The specimens should be made and cured in accordance with the procedure *Practice for Making and Curing Concrete Test Specimens in the Field (ASTM C31).* For laboratory work, the procedure *Method of Making and Curing Concrete Test Specimens in the Laboratory* (ASTM C192) may be used. The testing of the specimens should be done in accordance with ASTM C39, *Test Method for Compressive Strength of Cylindrical Concrete Specimens* and ASTM C78, *Test Method for Flexural Strength of Concrete (Using Simple Beam and Third-Point Loading).*

At times it is desirable to make compressive or flexural strength tests of specimens taken from the hardened concrete. The work should be done as specified in *Method of Obtaining and Testing Drilled Cores and Sawed Beams of Concrete* (ASTM C42).

STRENGTH TEST CYLINDERS (ASTM C31)

Test specimens are sampled from concrete being used in construction to determine if the concrete has the specified compressive strength. Field control specimens of various ages indicate the rate of strength gain and the effectiveness of job site curing. Great care must be exercised in the molding, protection, and curing of strength test cylinders.

The compressive strength field test requires that a sample of the concrete be taken at two or more regularly spaced intervals during discharge of the middle portion of the batch. Before the molds are filled, the individual portions of the sample should be remixed with a shovel to insure uniformity. The batch of concrete from which the samples are taken is noted as to its location in the work. In addition, the air temperature and any unusual conditions are noted.

The compressive test specimen is made in a watertight cylindrical mold. Standard cylindrical molds are 6 inches in diameter by 12 inches in length if the coarse aggregate does not exceed 2 inches in nominal size. The mold is filled in three equal layers. Each layer is rodded with 25 strokes of a 5/8-inch round steel rod about 24 inches long, with a hemispherical tip. When rodding the second and third layers, the rod should just break through into the underlying layer of concrete. Reinforcing rods or other tools should not be used as the rod. After the top layer is rodded, the surface of the concrete is struck off level with a trowel. The surface is immediately covered with a glass, metal, or plastic plate to prevent evaporation.

Molds should be placed on a rigid horizontal surface that will be free from vibration and other disturbances. During the first 24 hours, the test specimens should not be moved. They should be stored under conditions that will prevent the loss of moisture and maintain the temperature within a 60° to 80°F range. As soon as possible following the initial 24 ± 8 hour curing period, the cylinders are sent to a laboratory. Here the specimens are removed from the molds and protected from loss of moisture. Within 30 minutes standard curing is started at a temperature of 73.4 ± 3° F.

Standard curing procedures call for either laboratory curing or field curing. Laboratory curing gives a more accurate indication of the potential quality of the concrete. Field-cured specimens may give a more accurate interpretation of the actual strength in the structure or slab; however, these specimens offer no explanation of whether any lack of strength is due to errors in the proportioning of the ingredients, poor materials, or unfavorable curing conditions. On some jobs both methods are used, especially when the weather is unfavorable for the proper interpretation of the tests. The laboratory test result is the one that always prevails.

Concrete specimens of all ages should be protected from rough handling. In all cases, the specimens should be kept upright until they have hardened. If these precautions are ignored, the concrete test specimens will indicate lower than normal strengths. Rough handling and the lack of proper curing usually result in erratic and low-strength tests.

Laboratory compression tests are made in accordance with ASTM Designation C39, *Test Method for Compressive Strength of Cylindrical Concrete Specimens.* The specimen is placed in a testing machine where a load is applied and increased until the specimen fails. The compressive strength of the specimen then is calculated to the nearest 10 psi.

TEMPERATURE CONTROL

Changes in construction technology, greater understanding of plastic concrete, and new and specialized admixtures (see Table 8–7 Admixtures) have greatly increased the ability of concrete to be placed in adverse weather conditions. This has brought about ASTM C1064 *Test Method for Temperature of Freshly Mixed Portland Cement Concrete.* This test method provides a means for measuring the temperature of freshly mixed concrete. It may be used to verify conformance to a specified requirement for temperature of concrete.

EQUIVALENT ASTM AND AASHTO SPECIFICATIONS

See Table 9–1 for tabulation of equivalences.

STUDY/DISCUSSION QUESTIONS

1. What is the ASTM and why was it formed?
2. Explain why a slump test is used.
3. What does a change in slump indicate?
4. When should a concrete sample be taken for a slump test?
5. Explain the complete procedure for making a slump test.
6. After the cone is removed from a concrete specimen, how is the slump measured?
7. Of what value are compressive tests?
8. Explain the procedure for obtaining test cylinders.
9. How should the preliminary storage and curing of compressive test specimens be handled?
10. What precautions should be observed when handling and shipping compressive test specimens?
11. Name some factors affecting air content in concrete.
12. Explain how a simple field test is made to check possible changes in the air content of fresh concrete.

ASTM Designation	AASHTO Designation	ASTM Designation	AASHTO Designation
C 29-78	T 19-80	C 174-82	T 148-86
C 31-84*	T 23-86	C 183-83	T 127-85
C 39-83b	T 22-86	C 184-83	T 128-86
C 40-79*	T 21-81	C 185-80	T 137-82
C 42-84a	T 24-86	C 187-83	T 129-85
C 70-79*	T 142-81	C 188-84	T 133-86
C 78-84	T 97-86	C 190-82	T 132-84
C 85-66(1973)	T 178-83	C 191-82	T 131-85
C 87-69(1975)	T 71-80	C 192-81	T 126-86
C 88-76	T 104-86	C 204-84	T 153-86
C 109-84	T 106-86	C 231-82el	T 152-86
C 114-83a	T 105-85	C 232-71(1977)	T 158-74
C 115-79b	T 98-81	C 233-82	T 157-79
C 116-68(1980)	T 140-70	C 234-71(1977)	T 159-74
C 117-80*	T 11-85	C 266-77	T 154-82
C 123-83*	T 113-86	C 293-79	T 177-81
C 127-81*	T 85-85	C 301-79*	T 281-84
C 128-79*	T 84-86	C 305-82	T 162-84
C 131-81	T 96-83	C 359-83	T 185-86
C 138-81*	T 121-86	C 403-80	T 197-82
C 142-78*	T 112-81	C 430-83	T 192-86
C 143-78	T 119-82	C 451-83	T 186-86
C 151-84	T 107-86	C 496-71(1979)	T 198-74
C 156-80a	T 155-82	C 497/0497M-83	T 280-84
C 157-80	T 160-86	C 566-78	T 255-83
C 172-82	T 141-84	C 666-84	T 161-86
C 173-78*	T 196-80	C 702-80	T 248-83
		C 918-80	T 276-83

ASTM – American Society for Testing and Materials

AASHTO – American Association of State Highway and Transportation Officials

As an aid to the user of this volume, the following tabulation shows equivalences between ASTM and AASHTO specifications. Where an asterisk (*) follows the ASTM number, there is a difference between the two specifications.

Table 9–1 Tabulation of equivalences.

UNIT 10

READY-MIXED CONCRETE

Most concrete used in construction is mixed and delivered to the job site in a fresh, fluid state from a central plant remote from the construction location. Exceptions to this statement involve either massive structures such as dams, where it is economical to provide a concrete plant constructed exclusively for that particular job, or, in the opposite case, on a very small project requiring less than a full load of ready-mixed concrete.

PRODUCTION OF READY-MIXED CONCRETE

Ready-mix concrete plants use precision scales to weigh the ingredients and then batch portland cement with aggregates and water to achieve the desired concrete mixture.

The producer is responsible for selecting and proportioning the ingredients as specified by the purchaser and for delivering the concrete in good condition. After delivery, the quality of the hardened concrete depends entirely on the user who assumes responsibility for placing, finishing, and curing. Obtaining concrete construction of high quality is a joint responsibility of the producer and the user.

Ready-mixed concrete is manufactured for delivery to the job site in a plastic and unhardened state. There are three types of mixing, any one of which can be used.

1. *Transit-mixed* concrete is mixed completely in a truck mixer.
2. *Central-mixed* concrete is mixed completely in a stationary mixer and is delivered in a truck agitator; that is, a truck mixer operating at agitating speed, or in a special nonagitating truck.
3. *Shrink-mixed* concrete is mixed partially in a stationary mixer and then is completed in a truck mixer.

ASTM Designation C94, *Specification for Ready-Mixed Concrete,* requires that mixers and agitators be operated within the limits and speed of rotation specified by the equipment manufacturer. The agitating speed is about 2 to 6 rpm, and the mixing speed is about 8 to 12 rpm. Truck mixers consist essentially of a mixer with a separate water tank and water measuring device mounted in a chassis. Agitator trucks are similar except that they lack a water supply.

ASTM C94 requires that when a truck mixer is used either for complete mixing or for completing partial mixing, each batch of concrete should be mixed in the range of 70 to 100 revolutions of the drum or blades at the rate of rotation designated by the manufacturer as the *mixing speed.* Any additional mixing should be done at the speed designated by the manufacturer as the *agitation speed,* unless batch uniformity is shown to be inadequate, as specified in the standard. The specifications also require that concrete be delivered and discharged within 1 1/2 hours, or before the drum has revolved 300 times, after the introduction of the water to the cement and aggregates or the cement to the aggregates, unless otherwise specified by the purchaser.

ADVANTAGES OF READY-MIXED CONCRETE

There are a number of practical benefits offered by ready-mixed concrete. Because of these benefits, this industry has shown a phenomenal growth in recent years.

The use of modern precision batching equipment in the ready-mix industry assures purchasers of an accurately proportioned mix. The thorough mixing of each batch helps produce a concrete of uniform quality, durability, and high strength.

The use of ready-mixed concrete helps create a more efficient operation at the job site. This type of concrete eliminates the time spent in estimating the quantities of cement and aggregates required and the labor needed to handle and mix the materials. The problem of finding space at the job site for aggregate and cement storage is eliminated. There are no excess materials to clean up after a job is completed. Specialized equipment and personnel for job mixing are not required. In many cases, such equipment and personnel, where required, are used inefficiently because of fluctuating demand for concrete.

Virtually any construction location is accessible to a ready-mix truck. This feature makes it possible to deliver concrete to a specific job site at the time required, so that the concrete can be deposited immediately in the forms. Valuable construction crew time is not lost waiting to start or finish a job.

Efficient large volume production makes ready-mixed concrete an economical material for use in modern construction. It is delivered in the exact quantity desired, thus eliminating waste and costly reordering. It is not necessary to rent, lease, or purchase a mixer for use on the job. There is no guesswork about the cost of the concrete, since ready-mix producers will quote the price of their product before making delivery.

SPECIFICATIONS FOR READY-MIXED CONCRETE

Many engineers and architects now designate standard specifications for ready-mixed concrete. One specification commonly used is ASTM C94, *Specification for Ready-Mixed Concrete.*

The specification covers the producer's responsibility for ready-mixed concrete through delivery to the job site. Among the subjects covered are the basis of purchase, materials, quality of concrete, tolerances in slump, and measuring materials. The specification does not cover the placement, consolidation, curing, or protection of the concrete after delivery to the purchaser. These items are the responsibility of the user.

In the absence of designated applicable general specifications, the purchaser must select one of the three following alternatives for specifying the quality of the concrete.

Alternative No. 1 is used when the purchaser assumes responsibility for the design of the concrete mixture. The purchaser must specify the following:

(1) Cement content in pounds per cubic yard of concrete.

(2) Maximum allowable water content in pounds per cubic yard, including surface moisture, but excluding the water of absorption of the aggregates.

(3) If admixtures are required, the type, name, and amount to be used.

Alternative No. 2 is used when the purchaser requires the supplier to assume responsibility for the selection of the proportions for the concrete mixture. The purchaser must specify the following:

(1) Compressive strength of the concrete at the point of discharge. Unless otherwise specified, the strength refers to standard laboratory curing conditions tested at 28 days.

(2) If the complete mix design characteristics are desired by the purchaser, this information may be requested and received prior to the actual delivery of the concrete as:

- Type of cement and number of pounds used per cubic yard.
- Pounds of fly ash (pozzolans) used.
- Type and amount of water reducer used.
- Kind of accelerator, if used.
- Percent of air entrainment, if used.
- Aggregate used and type.
- Amount of water.
- List of other additives.

Cross reference these additives with those listed in Table 8–7 on admixtures. Most concrete suppliers are willing to discuss their recommendations; however, be cautious if your supplier offers a "secret" formula for ready-mixed concrete.

Alternative No. 3 is used when the purchaser requires the supplier to assume responsibility for the selection of the proportions for the concrete mixture after the purchaser indicates the minimum allowable cement content. The purchaser must specify the following:

(1) Same as Alternative No. 2.

(2) Minimum cement content in pounds per cubic yard.

(3) If admixtures are required, the type, name, and amount to be used.

REMIXING CONCRETE

The initial set of the concrete does not take place ordinarily for 2 or 3 hours after the cement is mixed with water. Fresh concrete that is left standing tends to dry out and stiffen somewhat before it begins to set. Such concrete may be used if it becomes sufficiently plastic when it is remixed that it can be completely compacted in the forms. Water should not be added indiscriminately to make the mixture more workable since it lowers the quality of the concrete just as the addition of a larger amount of water in the original mixing lowers the quality. A small amount of water may be added to delayed batches if the maximum allowable water-cement ratio is not exceeded and the concrete is remixed for at least half the minimum required mixing time.

PREPARING FOR READY-MIX DELIVERY

Ready-mixed concrete should be ordered well in advance for speedy and efficient service. The order should specify the time when the concrete is needed. This permits delivery schedules to be adapted to the schedule of the job. When an order is placed, the ready-mix producer will indicate approximately when the concrete will be delivered.

The placing operations must be planned far enough in advance of an order so that trucks can be scheduled to deliver concrete at the time it is needed. If the job is not ready for placing to begin when the trucks arrive, excessive mixing and hydration may cause the concrete to become too stiff to place and finish.

Ready-mixed concrete must be placed in the forms promptly, not more than about 1 1/2 hours after mixing. The allowable time between mixing and placing may be reduced during extremely hot weather. Adequate facilities and labor must be available to handle the concrete when it is delivered. Delays are costly in terms of time and quality lost. Delays may even result in the loss of the concrete. Expensive equipment may be tied up, thus affecting the cost per cubic yard of concrete. The purchaser may have to absorb this added expense.

The following figures illustrate some steps the purchaser of ready-mixed concrete can take to save time and reduce confusion on the job.

Fig. 10-1 Have forms well braced to keep them aligned while concrete is being placed. Oil the forms so they can be removed easily.

Fig. 10-2 Have wheelbarrows ready to speed the placing of the concrete immediately upon delivery at the work site.

Fig. 10-3 Have the subgrade dampened so the ground doesn't absorb water from the concrete. The subgrade must be level to insure uniform concrete thickness.

Fig. 10-4 Have strikeoff boards ready and screeds in place to insure that the concrete is finished to the proper grade.

Fig. 10-5 Have planks and runways in place to protect the existing construction and provide easy access for the trucks.

Fig. 10-6 Vibration equipment must be ready to insure proper compaction and bond to reinforcement and to help eliminate air pockets.

Fig. 10-7 Have straightedge ready to strike off concrete to the desired surface level before beginning the finishing operations.

Fig. 10-8 Curing materials must be ready to prevent the rapid evaporation of surface water as a first step in the proper curing procedure.

DANGERS OF ADDING WATER TO READY-MIXED CONCRETE

When the specified mix is ordered, the proper amount of water is determined at the plant and combined with the portland cement and aggregate. Since ready-mixed concrete is accurately designed for an individual job, the addition of extra water will seriously affect the strength, durability, and watertightness of the finished concrete.

Many producers discourage the use of high slump concrete by requiring that requests for the addition of extra water at the job site be noted on the delivery ticket and signed by the user to acknowledge the use of the extra water. Drivers are usually instructed not to add water indiscriminately. Concrete frequently may be delivered to the job "on the dry side." Water is then added to bring the slump to the specified amount. When this procedure is used, an adequate amount of mixing must be performed to incorporate the added water into the mix.

Many concrete workers fail to realize that too much water in a mix results in additional man-hours required for finishing. Finishers have to wait for the water to soak into the subgrade or evaporate. Stiff concrete usually is less expensive in terms of man-hours required. The initial placing labor may be more than for a soupy mix, but the concrete can be finished sooner.

AVAILABILITY OF READY MIX

Ready-mixed concrete is available for all types of construction. The mobility of mixer trucks and newer designs, such as 'over-the-cab' front end discharge mixers, make most job sites easily accessible for the delivery of ready-mixed concrete. If quality concrete is combined with good workmanship, the completed job will give years of dependable service.

STUDY/DISCUSSION QUESTIONS

1. Identify and explain the three methods by which ready-mixed concrete is mixed and delivered.
2. What are the producer's responsibilities with regard to the five construction rules of quality concrete?
3. What are the three responsibilities of the purchaser or user?
4. Explain some of the advantages of using ready-mixed concrete.
5. Explain why advance planning and preparation are recommended.
6. What advance plans should the user make?
7. Why is it never advisable to add water to ready-mixed concrete at the job?
8. Will excess water affect the finishing time? If so, how?
9. What can architects, engineers, and others use as a standard specification for ready-mixed concrete?

10. **Explain the five items that must be stated explicitly in a specification for ready-mixed concrete.**
11. Figures 10–1 through 10–8 indicate preparation for freshly delivered concrete. Name and discuss them.

UNIT 11

JOB MIXING CONCRETE

As stated in a previous unit, the purchaser of concrete specifies the mix desired when ordering ready-mixed concrete. The purchaser is also responsible for the concrete after it is delivered to the job. However, in the case of job-mixed concrete, the user is responsible for selecting good ingredients and using the proper mix.

The selection of a proper mix is based on the kind of work or exposure condition the concrete will be subjected to. The design of concrete mixtures involves the determination of the most economical and practical combination of ingredients to make the concrete workable in its plastic state and to make it develop the required qualities when hardened. Commercial laboratories perform this design service for many ready-mixed concrete plants as well as for specific projects. However, for small concrete jobs this service is not always economical or practical; therefore, someone on the job must determine the mix.

Quality concrete gives maximum service and satisfaction. The use of quality concrete results in durable, strong, and watertight installations. It is just as easy to make high quality concrete for excellent service as it is to make poor quality concrete.

OBJECTIVES OF A CONCRETE MIX

A properly designed concrete mix achieves three objectives: (1) workability of the fresh concrete, (2) required qualities of the hardened concrete, and (3) economy.

The workability of fresh concrete is the property that determines the amount of effort required to consolidate and finish the concrete. Although workability is difficult to measure, it can be judged readily by experienced concrete technicians.

Assuming that acceptable materials are used, the water-cement ratio determines the following qualities of hardened concrete: resistance to freezing and thawing, watertightness, wear resistance, and strength.

An economical concrete mix design should provide the maximum yield without sacrificing concrete quality. Since the quality depends mainly on the water-cement ratio, the water requirement should be minimized to reduce the cement requirement. Steps which can be taken to minimize the water and cement requirements include the use of (1) the stiffest practical mixture, (2) the largest aggregate size practical for the job, and (3) the optimum ratio of fine to coarse aggregate.

DETERMINING EXPOSURE CONDITIONS

The kind of work and the exposure conditions to which the concrete will be subjected should be known before a mixture is designed. There are three distinct classifications of exposure conditions.

Mild Exposure

Mild exposure conditions mean that the concrete is not subjected to abrasion or severe weather, such as in foundations and installations in mild climates where the temperatures rarely fall below freezing.

Normal Exposure

Under normal exposure conditions, the concrete is used for watertight structures and is subjected to abrasion or weather, such as in floors, sidewalks, paved barnyards, tilt-up wall panels, and tanks.

Severe Exposure

Severe exposure conditions mean that the concrete is subjected to severe wear, weather, weak acids or alkali solutions, such as in silos, mangers, and floors in dairy plants.

WATER-CEMENT RATIOS FOR EXPOSURE CONDITIONS

When mixing concrete on the job, the purpose of the concrete must be considered. The proper mix is based on the kind of work or exposure conditions to which the concrete will be subjected. As stated previously, the desirable properties of quality concrete depend on the quality of the water-cement paste. To insure quality concrete, the following desirable water-cement ratios were established in terms of the maximum water and minimum cement content for the three exposure conditions.

For Mild Exposures. The maximum water/cement ratio is 0.65, or 292.5 lb. of water to 450 lb. of cement per cubic yard of concrete.

Exposure: concrete not subject to abrasion or severe weather.

Example: foundations and installations in mild climates.

For Normal Exposures. The maximum water/cement ratio is 0.55, or 302.5 lb. of water to 550 lb. of cement per cubic yard of concrete.

Exposure: concrete for watertight structures and when subjected to abrasion and weather.

Example: floors, sidewalks, paved barnyards, tilt-up wall panels, and tanks.

For Severe Exposures. The maximum water/cement ratio is 0.45, or 292.5 lb. of water to 650 lb. of cement per cubic yard of concrete (air entrained).

Exposure: concrete is subjected to severe wear, weather, weak acids, or alkali solutions.

Example: silos, mangers, and floors in dairy plants.

The use of more or less aggregates (fine or coarse) varies the consistency of the mix to make it more workable for a specific job.

SLUMP

Under conditions of uniform operation, changes in slump indicate changes in materials, mix proportions, or water content. To avoid mixes that are too stiff or too fluid, slumps within the limits given in Table 11-1 are suggested. Concrete with a slump greater than 6 inches should not be used.

WATER-CEMENT RATIO BASED ON REQUIRED COMPRESSIVE STRENGTH

Under certain conditions, the water-cement ratio is selected on the basis of concrete strength. In such cases, it is recommended that tests be made with the job materials to determine the relationship between the water-cement ratio and the strength. If data for this relationship cannot be obtained because of time limitations, the water-cement ratio may be

Types of Construction	Slump, in. Maximum*	Minimum
Reinforced foundation walls and footings	3	1
Plain footings, caissons, and substructure walls	3	1
Beams and reinforced walls	4	1
Building columns	4	1
Pavements and slabs	3	1
Heavy mass concrete	3	1

*May be increased 1 inch for methods of consolidation other than vibration.

Table 11-1 Recommended slumps for various types of construction.

estimated from the graphs in figure 11-1. A majority of the results of strength tests made by many laboratories using a variety of materials appear within the strength-band curves.

SELECTING GOOD INGREDIENTS

Concrete ingredients consist of aggregates, water, and portland cement, plus an air-entraining agent if exposure conditions require it.

Aggregates

Aggregates such as sand, gravel, and crushed stone, are used in concrete to provide volume at low cost. The aggregates can be called a filler material since they total 60 to 80 percent of the volume of the concrete. The aggregates, both fine and coarse, must be clean, sound, and well graded. An important consideration for any job is the aggregate size. Generally, the most economical mix is obtained by using the largest aggregate that is practical for the specific job. Fine aggregate consists of particles ranging in size from 1/4 in. to a size small enough to pass through a sieve having 100 openings to the inch.

Mixing Water

The water used for concrete should be clean and free of acids, alkalies, oils, sulfates, and other harmful materials. A good rule for all jobs is that the water used for concrete should be fit to drink. If drinkable water is not available, tests should be made to determine the suitability of the water.

Portland Cement

When designing the mix, the portland cement selected must be the type best suited for the job. The cement should be free of hard lumps.

Air

Air present in the form of microscopic bubbles improves the durability of concrete. Air-entrained concrete should be used where the hardened concrete is exposed to alternate freezing and thawing cycles or where deicing chemicals are used. This type of concrete is recommended for almost all outside concrete work. Because air-entrained concrete improves

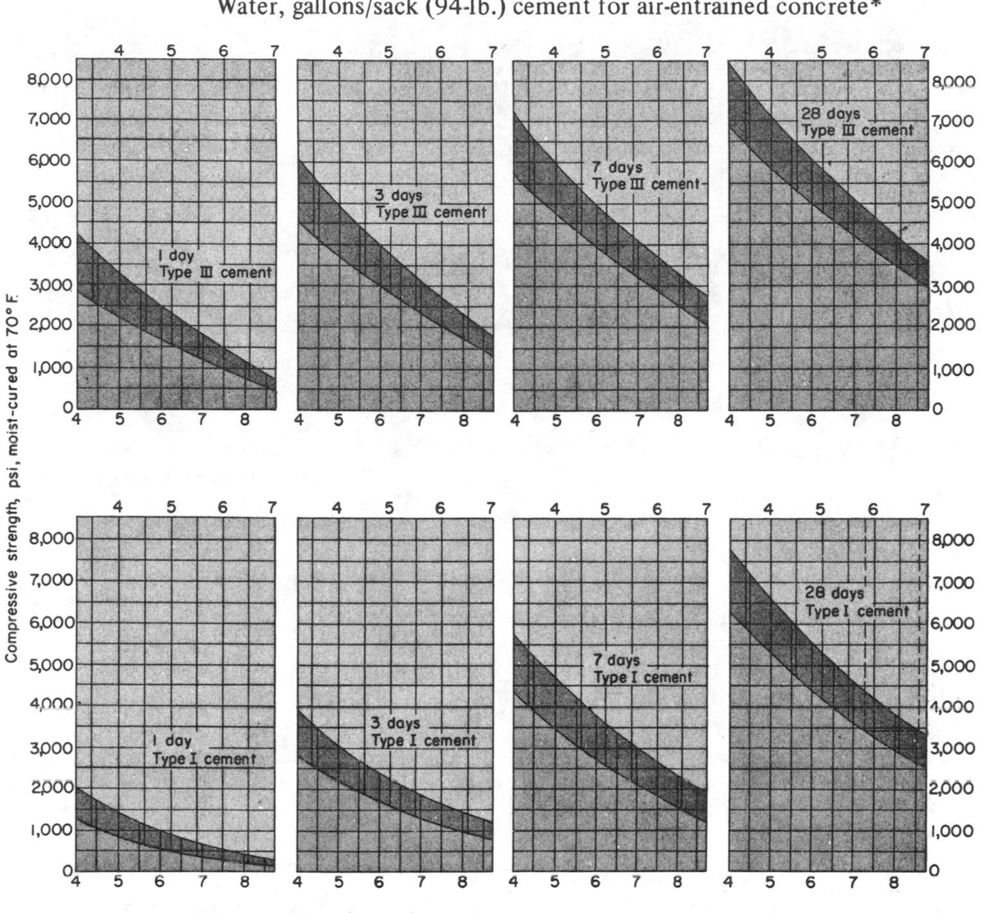

*Concrete with air content within recommended limits and maximum aggregate size of 2 inches or less.

Fig. 11-1 Relationship between the water-cement ratio and the compressive strength for Types I and III portland cements at different ages. These relationships are approximate and should be used only as a guide in the absence of data on job materials.

workability, it should also be considered for most indoor work. To entrain air, either an air-entraining portland cement or an air-entraining admixture must be used. Unless an air-entraining cement is used, enough admixture must be added to a portion of the mixing water to produce the desired air content. The amount recommended by the manufacturer produces the desired air content in most cases.

ALLOWING FOR MOISTURE IN SAND

Once the proper water-cement ratio for the kind of work or exposure condition is selected, the effect of moisture in the sand (aggregate) must be considered.

It is not enough to know the gross amount of water used. It is necessary to consider the amount of water carried by the fine aggregate. The water carried by the coarse aggregate is generally so small that it can be ignored.

It is important to determine the net amount of water to be used with sands having various degrees of wetness. Excess moisture is free to react with the cement and thus change

Fig. 11-2 If sand falls apart it is damp.

Fig. 11-3 If sand forms a ball it is wet.

Fig. 11-4 If sand sparkles and wets the hand, it is very wet.

the water-cement ratio. As a result, the sand must be tested for its water content and an allowance made.

Dry sand is seldom available for concrete work. Sand is considered to be dry if it is spread out in a thin layer and dried in the sun or warm air.

Most sand used on the job contains a surprising amount of water. When used in small jobs, it is possible to estimate the amount of water in the sand by squeezing it in the hand.

Damp sand feels slightly damp to the touch but leaves little moisture on the hands, figure 11-2. Damp sand usually contains about 1/4 gallon of water per cubic foot of sand (about 2 percent by weight).

Wet sand, which is the kind usually available, feels wet and leaves a little moisture on the hands, figure 11-3. Wet sand contains approximately 1/2 gallon of water per cubic foot of sand (about 4 percent by weight).

Very wet sand is dripping wet when delivered on the job and leaves more moisture on the hand than wet sand, figure 11-4. Very wet sand contains about 3/4 gallon of water per cubic foot of sand. When the sand contains large amounts of fine material, it may have as much as 1 1/3 gallons of water per cubic foot (from 6 to 10 percent by weight).

PROPORTIONING THE MATERIALS

After the sand is classified as damp, wet, or very wet, Table 11-2, page 68, may be used to determine the trial proportions of the ingredients for any particular job.

Using the Maximum Aggregate Size Permissible

As the size of the coarse aggregate increases, less paste is necessary. As a result, more concrete is obtained per hundredweight of cement used. This means that the most economical mix generally is obtained by using the largest size of coarse aggregate practical for the job.

In general, the maximum aggregate size depends on the size and shape of the concrete members and the amount and distribution of the reinforcing steel. The maximum aggregate size usually should not exceed:

- One-fifth the dimension of the nonreinforced member, or

	Gal. of water added to 1-cwt. batch if sand is:			Suggested mixture for 1-cwt. trial batches*		
	Damp	Wet (average sand)	Very Wet	Cement cu. ft.	Aggregates	
					Fine cu. ft.	Coarse cu. ft.
For Mild Exposure 0.65 water/cement ratio	For 1 1/2-in. max. aggregate size 6 1/4	5 1/2	4 3/4	1	3	4
For Normal Exposure 0.55 water/cement ratio	For 1-in. max aggregate size 5 1/2	5	4 1/4	1	2 1/4	3
For Severe Exposure 0.45 water/cement ratio	For 1-in. max. aggregate size 4 1/2	4	3 1/2	1	2	2 1/4

*Mix proportions will vary slightly depending on the gradation of aggregate sizes. (1 cwt. of cement is approximately equal to 1 cu. ft., loose volume.)

Table 11-2 Trial mix proportions.

- Three-fourths the clear spacing between the reinforcing bars or between the reinforcing bars and forms, or
- One-third the depth of nonreinforced slabs on grade.

PREPARING TO MIX CONCRETE

To reduce the labor on a job, and thus save money, the concrete mixing site should be as close as possible to the construction location. The size of the mixer used depends upon the size and type of the job and the manpower available. Concrete for large flat jobs is easy to place and a large mixer will help to speed the work. The capacity of the mixer should be checked before the ingredients are added; this information is usually given on a plate attached to the mixer.

The aggregates should be placed so that they are convenient to the mixer operators. The cement must be stored in a dry location. If it is stored outside, it should be stored on a platform raised off the ground and covered with a tarpaulin, polyethylene sheeting, or other suitable cover. Ample water and measuring containers should be available.

Measure all Materials

All materials must be measured accurately. The water can be measured in a pail marked on the inside to show quarts and gallons. On small jobs, a pail may be used to measure the cement, sand, and gravel. In mixing 1/2- or 1-bag batches, remember that 1 cwt. of cement is about 1 cu. ft. in loose volume. Aggregates are conveniently measured in bottomless boxes made to hold exactly 1 cu. ft., 2 cu. ft., or whatever volume is required.

MIXING THE CONCRETE

All concrete should be thoroughly mixed until it is uniform in appearance and all materials are uniformly distributed. This means that concrete should be mixed for at least

1 minute and preferably for 3 minutes after all of the materials are placed in the mixer. There is little benefit to be gained in mixing for more than 3 minutes.

Under normal mixing conditions, about 10 percent of the mixing water should be placed in the mixing drum before the dry materials are added. Water then is added uniformly with the dry materials until about 10 percent of the water remains. This amount is added after all of the other materials are in the drum.

The accurate measurement of all materials, including water, is necessary to insure the production of uniform batches of quality concrete. Uniformity is assured by using accurate weighing or measuring equipment. Good judgment is necessary to secure concrete of the proper consistency. Concrete in the mixer usually appears to be drier than it does after being placed.

Mixers used on the job vary in drum capacity from 1/2 cu. ft. to 2 cu. ft., or larger, Regardless of size, the mixers should not be loaded above their rated capacity. Mixers should be operated at the speeds for which they are designed. If the blades of the mixer become worn or coated with hardened concrete, the mixing action is less efficient. Badly worn blades should be replaced and hardened concrete should be removed before each new run of concrete is mixed.

CLEANING EQUIPMENT

The mixer and other equipment or tools which come into contact with the concrete must be cleaned when necessary during use and must be thoroughly cleaned after use.

Mud and dirt on the frame and mixer drum usually can be hosed off with water. Periodically, it may be necessary to clean the cement film from the exterior of the mixer. A weak vinegar solution (10 percent acetic acid) can be used for this purpose. Areas cleaned with acid solutions should be rewaxed.

Concrete buildup inside the drum must be prevented to achieve the greatest efficiency in mixing and discharging. The interior of the drum should be hosed thoroughly with water. However, residues of hardened cement and concrete may remain. It may be necessary to scrape or brush the interior vigorously. One of the easiest ways to damage a drum and mixing blades is the continuous use of a heavy hammer to remove the concrete buildup. A recommended method for loosening stubborn material is to use a solution of 3 parts of water to 1 part of muriatic acid. The solution is applied and allowed to penetrate for about 30 minutes before hard rubbing or brushing is resumed.

Commercially available releasing agents are satisfactory and reduce the man-hours required to clean mixers and concrete equipment and tools. Solvents are also available to protect equipment and tools from cement dust, concrete, and dirt.

STUDY/DISCUSSION QUESTIONS

1. What is the objective of a properly designed concrete mixture?
2. Explain how the most economical mix can be obtained for a given exposure condition.
3. What are the proper mixes (water-cement ratios) for mild exposure conditions? Normal exposure conditions? Severe exposure conditions?

4. Why is it necessary to consider the moisture content of the fine aggregate?
5. Explain how a sand is classified as dry, damp, wet, or very wet.
6. How much moisture does wet sand contain?
7. Explain what the phrase "select good ingredients" means.
8. Why is it advisable to use the largest aggregate size practical for a job?
9. What should be considered when determining the mixing site?
10. Where should the aggregates be located in relation to the mixer?
11. What effect does the quantity of mixing water have on the quality of concrete?
12. What is the volumc of 1 cwt. of cement?
13. Explain the procedure for loading a mixer.
14. Explain how a mixer should be cleaned after use.

UNIT 12

TOOLS FOR PLACING AND FINISHING CONCRETE SLABS

The large variety of concrete working tools available, both old and new, makes it impossible to describe them all in just one unit. Therefore, this unit deals only with those tools that are most commonly used for placing and finishing concrete.

STRAIGHTEDGE

A straightedge or strike-off board, figure 12-1, is the first finishing tool used after the concrete is placed. The straightedge is used to level the concrete surface to the proper grade.

A straightedge may be made of metal or of lumber. The tool must have sufficient rigidity to perform the strike-off operation. The wood straightedge is generally made of 1 1/4-inch stock. The straightedge may have a shoe strip attached at the bottom as shown in figure 12-2. A magnesium strike-off board is usually 1 inch thick, 4 1/2 inches wide, and from 6 to 16 feet in length.

It is preferred practice to use a tool specifically designed for striking off concrete. However, if a tool is not available, a straight piece of 2″ x 4″ or 2″ x 6″ lumber may be used.

The striking surface of the straightedge should be straight and true. The straightedge should be longer than the widest distance between the screeds or edge forms. The maximum span that should be struck off by hand is approximately 12 feet. With larger spans, there is a tendency for a sag or rollback to develop. When this occurs, power equipment should be used or the screeds should be placed closer together.

Fig. 12-1 Straightedge or strike-off board.

Fig. 12-2 Straightedge with shoe strip.

DARBY

A darby, figure 12-3, is a long, flat, rectangular piece of wood, aluminum, or magnesium. The darby is from 3 to 4 inches wide and has a handle on the top. This tool is used to

Fig. 12-3 Darby.

float the surface of the concrete slab immediately after it is struck off to prepare it for the next step in finishing. This tool is used to eliminate any high or low spots or ridges left by the straightedge. The use of the darby should also embed the coarse aggregate sufficiently for subsequent floating and troweling.

BULL FLOAT

A bull float, figure 12-4, is a large flat, rectangular piece of wood, aluminum, or magnesium. This tool is usually 8 inches wide and 42 to 60 inches long with a handle 4 to 16 feet long. The function of the bull float is essentially the same as that of the darby. However, the bull float enables the concrete finisher to float a larger section of a wide slab. The bull float is more commonly used outdoors or in locations where there is enough room to use the long handle.

EDGER

Edgers, figure 12-5, are available in many sizes; however, they all are about 6 inches long and vary in width from 1 1/2 to 4 inches. The lip may range in size from 1/8 to 5/8 inch. The radii of the lip ranges from 1/8 to 1 1/2 inches. The curved-end edger is com-

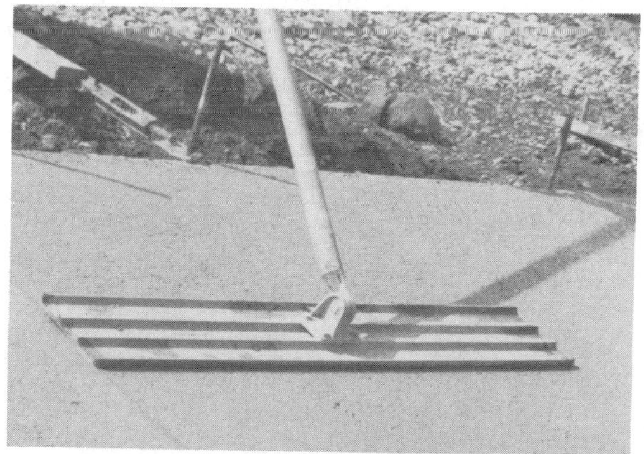

Fig. 12-4a Bull float.

Fig. 12-4b Floating concrete.

Fig. 12-5 Edger.

Fig. 12-6 Jointer or groover.

monly used. Edgers are used to produce a radius at the edge of a slab to improve the appearance and reduce the risk of damage to the edge.

JOINTER OR GROOVER

Jointers or groovers, figure 12-6, are about 6 inches long and vary from 2 to 4 1/2 inches in width. These tools have shallow, medium, or deep bits (cutting edges) running from 3/16 to 1 inch in depth. The jointer is used to cut a joint partly through fresh concrete. The purpose of the joint may be to improve the appearance of the work or to act as a control joint if cut 1/4 to 1/5 the depth of the slab. The control joints (also called dummy or contraction joints) are used to predetermine the location of any possible cracks.

POWER JOINT CUTTER

Another method of cutting joints in concrete slabs is to use an electric or gasoline-driven saw, figure 12-7, fitted with a shatter-proof abrasive or diamond blade. A power cutter produces a narrow joint that minimizes the possibilities of spalling at the joint due to traffic. In general, the joint is cut in the concrete surface 4 to 12 hours after the concrete has hardened, or as soon as the concrete surface will not be damaged by the saw kerfs.

Fig. 12-7 Power joint cutter.

HAND AND POWER FLOATS

Hand floats are made of aluminum, magnesium, or wood. Aluminum or magnesium floats, figure 12-8, are usually made in two sizes, either 12 or 16 inches long by 3 1/2 inches

Fig. 12-8a Lightweight metal float.

Fig. 12-8b Hand float.

wide. Wood floats, figure 12-9, are available in 12-, 15-, or 18-inch lengths and 3 1/2- or 4 1/2-inch widths.

Fig. 12-9 Wood float.

The hand float is used to prepare the concrete surface for troweling. Hand floats also are used to float the concrete around pipes and columns, against walls or other areas where power floats cannot be used.

After the concrete is stiff enough to support the weight of a man, a power float can be used to compact the concrete, figure 12-10. These floating machines perform the same operations as hand floating but cover larger areas faster.

Power floats are driven by electric motors or gasoline engines. A power float, with a rotating disk about 2 feet in diameter, is used only on no-slump concrete for heavy duty floor construction. For slab construction, a power trowel is used, figure 12-12, with clip-on or shoe-type float blades.

HAND AND POWER TROWELS

The steel hand trowel is available in a variety of sizes ranging from 10 to 20 inches in length and 3 to 4 3/4 inches in width, figure 12-11.

Each size of hand trowel has a specific use. For the first troweling of concrete flatwork, a wide trowel 16 to 20 inches long is generally used. For the final troweling of the slab, most finishers prefer a *fanning* trowel 14 to 16 inches long and 3 to 4 inches wide.

A power trowel has three or four rotating steel trowel blades, figure 12-12. The blades are powered by electricity or by a gasoline engine.

The use of steel trowels gives the concrete surface a dense, smooth finish.

Fig. 12-10 Power float.

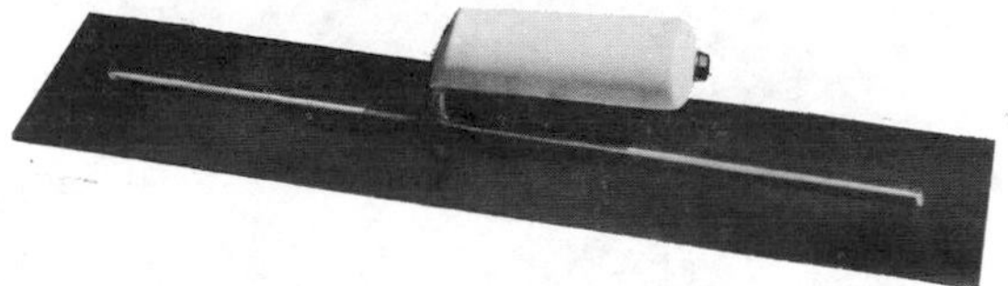

Fig. 12-11 Steel hand trowel.

Fig. 12-12 Power trowel.

STUDY/DISCUSSION QUESTIONS

1. Name some of the tools commonly used for finishing concrete.
2. Is it necessary that a straightedge be made of wood? Why?
3. When is a bull float used?
4. Why are jointers used?
5. When should a power cutter be used for cutting joints?
6. What are some of the materials used to make floats?
7. Explain the advantages of power floats or trowels.

UNIT 13

PLACING CONCRETE

The preparations to be completed prior to the placing of concrete include compaction, trimming, moistening the subgrade, erecting the forms, and setting the reinforcing steel.

The time element, the transportation of the concrete on the job, and the various methods of placing the concrete in the forms are probably the most important phases of placing concrete.

TIME

The concrete must be delivered to the job site and discharged from a truck within 1 1/2 hours after the water is added to the mixture. In hot weather, or in conditions which contribute to the quick stiffening of the concrete, the allowable time is less than 1 1/2 hours. Usually, it is good practice to place the concrete in the forms within 30 to 60 minutes after mixing. This means that the site must be ready so that the ready-mixed concrete trucks are not kept waiting. If the truck must wait, then excessive mixing and hydration may cause the concrete to become stiff and difficult to place and finish. In addition, the next delivery will be behind schedule.

TRANSPORTATION ON THE JOB

The various methods used to move the concrete from the mixer or truck to the forms depend largely upon the job conditions. On small jobs, wheelbarrows are the usual means of transportation. However, concrete can be handled and transported by many methods, including the use of chutes, buggies operated over runways, buckets handled by cranes or cable ways, small rail cars, trucks, pumps to force the concrete through pipelines, and equipment to force the concrete through hoses pneumatically.

Regardless of the kind of transportation used, care must be exercised to prevent segregation as the concrete is being moved. A rather stiff mix and smooth runways are usually required to prevent segregation. However, the method of handling and transporting the concrete and the equipment used should not place a restriction on the consistency of the concrete. This factor should be governed by the placing conditions. If these conditions permit the use of a stiff concrete mix, then the equipment should be selected and arranged to ease the handling and transportation of such a mix.

The job must be planned so that the ready-mix truck can get as close to the job site as possible. This will save the time and effort required to move the concrete any distance by hand methods. Any debris and unnecessary equipment must be removed to give the ready-mixed concrete truck more freedom of movement.

Many methods are used to transport and handle the concrete from the batching plant to the job site. Some concrete jobs are large and require many pieces of special equipment; other jobs are smaller and can be completed using simple, common equipment.

The variety of equipment used to handle fresh concrete is increasing rapidly. This equipment includes hoppers and chutes, wheelbarrows and buggies, cranes and other hoisting equipment, conveyors, pumps, and helicopters. The main consideration in selecting the type of equipment to be used is an economic one; however, certain jobs require specialized equipment and thus the cost is a secondary consideration.

The segregation of aggregates, loss of entrained air, loss of cement paste, change in slump, and accumulation of harmful materials are only a few of the factors that must be considered when selecting the transporting and handling equipment.

The use of pumps is one of the most satisfactory methods of moving concrete from the truck discharge point to its final destination. In this handling method, the concrete is pumped under pressure through a system of rigid or flexible pipes or hoses. Pumped concrete is used for small jobs, such as a patio for a house. It is also used for larger, continuous production jobs that require the concrete to travel as high as 500 feet vertically and more than 1,000 feet horizontally.

Three basic types of pumps are commonly used: the piston displacement pump, the pneumatic displacement pump, and the squeezing action pump.

Several factors affect the pumpability of the concrete. These factors include the gradation of the aggregate (especially the fine aggregate), the cement content, and the air content of the mix. The mix design should be prepared under laboratory conditions according to accepted practices. Special consideration must be given to the final use of the concrete, the slump, aggregate gradation, and the water-cement ratio. Local aggregates can be an important part of the mix design. A field trial must be run prior to the final selection of any mix. Lightweight and structural lightweight aggregate concrete can be pumped successfully if the proper procedures are followed.

PLACING IN FORMS

All concrete forms must be clean, tight, adequately braced, and constructed of materials that will impart the desired texture to the finished concrete. Sawdust, nails, and other debris should be removed from the forms before the concrete is placed. Wood forms should be moistened before the concrete is placed, otherwise they will absorb water from the concrete and swell. In addition, the forms should be oiled or lacquered to make form removal easier.

Reinforcing steel should be clean and free of loose rust or mill scale at the time the concrete is placed. Any coatings of hardened mortar should be removed from the steel.

The concrete should be placed between the forms or screeds as close as possible to its final position. To consolidate the concrete, it should be mechanically vibrated or spaded as it goes into the form. Then the concrete is thoroughly spaded next to the forms to eliminate voids or honeycombing at the sides. In inaccessible areas, the forms can be tapped lightly with a hammer to achieve consolidation. This operation makes a dense concrete surface by forcing the coarse aggregate away from the form or face. The concrete should not be overworked while it is still plastic. Overworking will cause too much water and fine material to be brought to the surface. This may later lead to scaling or dusting.

In slab construction, the placing of the concrete is started around the perimeter at the far end of the work so that each batch is dumped against previously placed concrete. The concrete should not be dumped in separate piles and the piles then leveled and worked together. The concrete should not be placed in large quantities (big piles) and then allowed to run or be worked over a long distance to its final position. Honeycombing and segregation may occur as the mortar tends to flow ahead of the coarser material. Water should be prevented from collecting at the ends and corners of forms and along form faces.

If the concrete is placed along a form in layers not more than 12 to 18 inches thick, good uniformity will be obtained. Each layer should be thoroughly consolidated before the next layer is placed.

For some construction work, it is necessary to move the concrete horizontally within forms, such as beneath openings in walls. The horizontal distance should be minimized as much as possible.

When concrete is placed in tall forms at a fairly rapid rate there may be some bleeding of the water to the top surface, especially with non-air-entrained concrete. Bleeding can be reduced by placing the concrete at a slower rate and by using concrete with a stiffer consistency. When possible, the concrete should be placed to a level about a foot below the top of high forms. An hour or so should be allowed for the concrete to settle. The placing of the concrete should be resumed before setting occurs to avoid formation of cold joints. It is good practice to overfill the form by several inches and then cut off the excess concrete after it is partially stiffened.

The old rule that "good concrete is placed, not poured" is still very true. All concrete workers should remember this rule. The concrete should be placed where it is needed. It should not be pushed or dragged into place or be allowed to flow into place. When using buggies or wheelbarrows, the concrete should be dumped into the face of previously placed concrete, not away from the face.

DO NOT ADD EXCESS WATER AT THE JOB SITE

Ready-mixed concrete is usually delivered after it is accurately designed for a specific job. Addition of water will seriously affect all of the desirable properties of the finished concrete such as durability, strength, and watertightness. Excessive water should never be added so that the concrete will flow into forms without working it. Good concrete cannot be placed without some work.

USING A DROP CHUTE

The use of a chute is recommended when fresh concrete must be dropped more than 3 or 4 feet. If concrete is allowed to strike and bounce off one side of the form, it will separate and show stone pockets and sand streaking. Drop chutes are also used to prevent the buildup of dried mortar on reinforcements and forms. For thin sections, drop chutes made of rubber or metal should be used. For narrow wall forms, rectangular metal drop chutes are fabricated to fit between the reinforcing steel. Drop chutes are available in several lengths or in sections which can be hooked together so that the length can be adjusted as concreting progresses. Figure 13-1 shows a drop chute with a hopper.

CONSOLIDATING CONCRETE IN FORMS

The most common methods of consolidating concrete in forms are hand spading and vibration. The proper use of vibrators makes possible the placement of stiff, harsh concrete

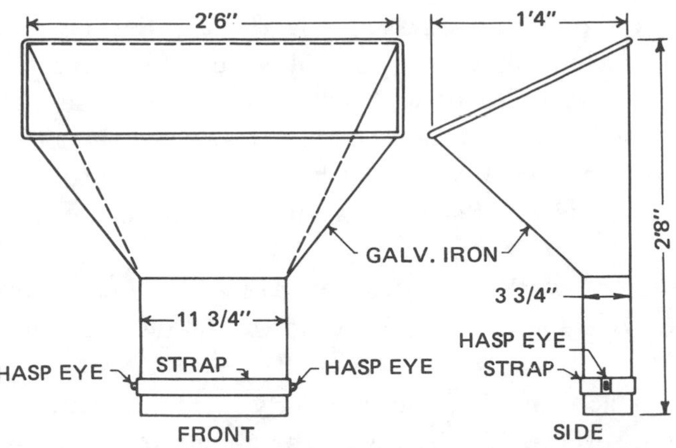

Fig. 13-1 A rectangular drop chute with hopper at top for placing concrete in narrow walls.

Fig. 13-2 Use of crane and bucket to place concrete. Note use of drop chute.
(Courtesy Portland Cement Association)

mixes that cannot be placed and consolidated readily by hand. Mixes that can be consolidated readily bv hand should not be vibrated as they are likely to segregate. In many cases, the slump can be less than one-half that required for hand placing, and in most cases, it can be reduced at least one-third by vibrating the mix.

Vibrators

One of the most important skills to be mastered in the placement of concrete against vertical surfaces is the consolidation of the concrete. This is accomplished by external and internal vibrators, spuds, and other consolidating devices. For internal vibration, the primary mechanical requirements for a specific vibrator are its size, power, frequency, and amplitude. The frequency is the number of times the vibrator head moves from side to side; the distance the vibrator head moves is the amplitude. Research indicates that as an active vibrator is placed into concrete, large aggregate particles are driven away from the vibrating head by contact. Smaller particles and mortar then flow to fill the spaces left by the large particles. As the larger, then smaller, particles reach the surface of the form, air voids and bubbles move up to the surface and escape. It is important, therefore, that the operator keep the vibrator head moving in an up-and-down direction to consolidate the concrete in lifts and to prevent honeycombing and air voids at the surface of the form.

Immersion-type vibrators, commonly called *spud vibrators*, are excellent tools to consolidate fresh concrete in walls or other formed work. The spud vibrator, a metal tubelike device, vibrates at the rate of several thousand cycles per minute. It is usually powered by gasoline, air, or electricity. When inserted in concrete for 5 to 15 seconds, the spud vibrator consolidates and improves the surface closest to the forms.

Vibrators are ideal for placing dry mixes. However, vibrators should be handled carefully in wet mixes so that too much water and fine sand will not be brought to the surface of the concrete. Systematic vibration will consolidate the concrete adequately. However, vibrators should never be overused. Vibrators should not be used to move concrete horizontally since this can result in segregation. If a vibrator is left in the concrete after the paste accumulates over the head, nonuniformity results. The length of time that a vibrator can be left in the concrete depends upon the slump of the concrete. Concrete with a high slump requires little or no vibration. Vibrator heads should never be held against reinforcement. Too much vibration, or using the vibrator haphazardly, can spread or destroy forms.

Form vibrators may be attached to the exterior of the forms. These vibrators are especially useful for consolidating concrete in thin-walled members and in metal forms.

STUDY/DISCUSSION QUESTIONS

1. How much time is permitted between the addition of water and concrete placement?
2. Describe some of the things that can happen when a ready-mix truck must wait to discharge a load of concrete.
3. Why should a ready-mix truck get as close to a job as possible?
4. What common methods are used to transport concrete at the job site?
5. What precautions should be observed in transporting concrete?

6. Is it advisable to flow or drag concrete? Why?
7. Explain how to insure consolidation of the concrete after it is placed in forms.
8. How can honeycombing and sand streaking be prevented?
9. How deep should the layers of concrete be when placed in forms?
10. When is it advisable to use drop chutes?
11. How long should concrete be vibrated to insure consolidation?
12. What happens if the concrete is vibrated for too long a period?
13. What is external vibration of concrete?

UNIT 14

FINISHING CONCRETE SLABS

The finish of the concrete depends on the final use of the completed job and the effect desired. Various colors and textures, such as exposed aggregates, may be used. Some surfaces are finished by striking them off to the proper contour and elevation. For other surfaces, a broom, float, or trowel finish may be specified. To finish concrete properly, it is essential to know the specific operations required.

CONSOLIDATING

Concrete must be consolidated to achieve a uniform, plastic mass. Consolidation eliminates stone pockets and large air voids. While many methods are used to consolidate concrete, the most common method is the use of a strike-off board or straightedge.

Other tools for consolidating include tampers, rollers or roller screed, and the vibrating screed.

Tampers should be used sparingly and only when consolidation cannot be accomplished readily by hand; otherwise, the concrete will segregate. For example, concrete having a low slump can be tamped.

STRIKING OFF

Striking off, also called screeding, is a leveling operation that removes humps and hollows and gives a true, even surface to the concrete. In many cases, the process of consolidation and strike off are combined into a single operation by the use of the strike-off board or straightedge. While the use of an actual tool is recommended, in an emergency, a 2 x 4 or 2 x 6 piece of lumber that is 1 to 2 feet longer than the section being finished may be used. The surface is struck off by moving the straightedge back and forth with a sawlike motion across the top of the forms or screed. The straightedge is advanced a short distance with each movement. A small amount of concrete is always kept ahead of the straightedge to fill in the low spots and maintain a plane surface. Vibrator screeds, roller screeds, and mechanical screeds are also used for striking off the concrete surface.

Immediately after the screeding process, the concrete must be darbied or bull floated before any free water bleeds to the surface. The bull float is frequently used because it has a long handle and is easy to use on wide slabs.

The use of the darby and the bull float eliminate the ridges and voids left by the strike-off operation. In addition, the bull-floating operation helps push large aggregate particles beneath the surface of the concrete and brings sufficient mortar to the surface to prepare for other finishing operations. Care must be taken during this operation to avoid overworking the surface.

The concrete may be left without further finishing. For most jobs, however, strike off is followed by one or more operations.

EDGING AND JOINTING

When the bleed water is gone, the water sheen (the glossy appearance) has disappeared from the concrete surface, and the concrete has started to stiffen, it is time for the remaining finishing operations. All open edges should be rounded off with an edger to prevent spalling. The edger should be moved back and forth until a finished edge is produced. The concrete finisher should insure that all of the coarse aggregate particles are covered and that the edger does not leave too deep an impression in the top of the slab. If the indentation is too deep, it may be difficult to remove in subsequent finishing operations. The initial edging should be done with a wide flange tool having a radius of 1/8 inch or less. Joints with greater radii are difficult to maintain.

The marks left by the edger and jointer should be removed by floating unless these marks are desired for decorative purposes. In these cases, the edger or jointer marks should be redone after the floating operation. If necessary, tooled joints and edges should be redone before and after troweling to maintain uniformity or to remove kinks.

Concrete expands and contracts slightly due to temperature changes. In addition, concrete may shrink as it hardens. Control joints, also called dummy or contraction joints, are cut across each concrete slab to control cracking. These joints do not extend completely through the slab, but are cut to a depth equal to 1/4 to 1/5 the slab thickness. The slab is weakened at these control joints. If the concrete cracks due to shrinkage or thermal changes, the crack usually occurs at the weakened control joint.

The control joints should be cut soon after the concrete is placed so that the larger pieces of coarse aggregate can be worked away from the joint. A simple method of cutting this type of joint is to lay a board across the fresh concrete and guide a groover along the back to cut the joint to the proper depth. The groover used should have a 3/4- or 1-inch bit. A concrete finisher may decide to use a brick mason's trowel to cut the joint to a depth of approximately 1/4 to 1/5 the slab depth. A groover is then used to finish the joint. In general, control joints are placed 10 to 15 feet apart on floor slabs, driveways, and feeding floors, and 4 to 5 feet apart on sidewalks.

The sawing of control (contraction) joints is commonly done using special concrete saw blades. Joints should be sawed 4 to 12 hours after the concrete is placed and finished. A sawed joint is clean and attractive and works well when cut to the proper depth. For large jobs, a special mobile concrete saw is used. On smaller jobs, such as sidewalks and driveways, a portable electric handsaw may be used. The operator must be careful to keep the saw blade straight so that it does not shatter.

Fig. 14-1 Sawing hardened concrete.

FLOATING

It is difficult to set a definite time to begin the floating procedure because of several variables such as concrete temperature, air temperature, relative humidity, and wind. The art of knowing just when to begin comes only through a great deal of on-the-job experience.

After the concrete is darbied or bull floated, it should be allowed to harden enough so that the weight of a man standing on it leaves only a slight imprint. Floating should not begin until the water sheen disappears. The surface should then be floated using wood, cork, or metal floats, or a finishing machine.

Aluminum or magnesium floats are recommended for working air-entrained concrete. These metal floats reduce the amount of work required by the finisher due to the fact that drag is reduced and the float slides over the concrete surface with a good floating action. Wooden floats tend to stick and tear the concrete surface.

Floating removes slight imperfections and fills small hollows in the surface of the concrete. This procedure also helps level and compact the concrete, embed large pieces of aggregate beneath the surface, and bring sufficient mortar to the surface to prepare it for other finishing operations.

FINAL FINISHING

A final finish is applied to concrete surfaces to give the desired appearance and texture. The final finishing is usually done immediately after floating, but must not be done until the surface is quite stiff yet workable.

BROOM FINISH

Some concrete projects such as driveways, walks, feeding floors, and lots and ramps for livestock require a coarse, scored surface. A rough texture can be made in the concrete surface by running a stiff broom crosswise (perpendicular) to the direction of travel. Almost any degree of coarseness or scoring can be obtained by using various types of brooms and amounts of pressure.

BURLAP OR BELT FINISH

A final finish on driveways, pavements, and similar types of construction can be made with a strip of burlap, canvas, or rubber from 6 to 12 inches wide. This material is used immediately after the wood float and is moved back and forth in a sideways direction as it is advanced. Burlap or belt finishing is performed in two operations. Initial 12-inch strokes are followed by a faster forward movement with 4-inch strokes.

HAND FLOAT FINISH

This finish is not to be confused with the initial floating operation. A second floating is necessary after the concrete has partially hardened so that a desired permanent finish can be produced. The hand float is used to produce gritty, nonslippery surfaces that wear well, are attractive, and provide good footing.

TROWELING

When a smooth, dense surface is desired, floating is followed by steel troweling. This operation should be delayed as long as possible so that the concrete is not too soft and plastic. Excessive troweling at this stage may cause crazing and may result in a surface having less wear resistance. Too long a delay, however, will result in a surface that is too hard to finish with a trowel.

It is not recommended that dry cement be spread on a wet surface to take up excess water. Such wet spots should be avoided where possible by adjusting the grading, mix proportions, and consistency. When wet spots do occur, finishing operations should be delayed until the water either disappears or is removed with a squeegee.

The first steel troweling should be sufficient to produce a smooth surface free of defects. Whether the troweling is done by hand or by power, the trowel blade must be kept as flat as possible against the surface. A trowel blade which is tilted or pitched at too great an angle will result in an objectionable "chatter" or "washboard" surface. It is not recommended that the first troweling be done with a new trowel. An older, used trowel is preferred since it can be worked quite flat without having the edges dig into the concrete.

A second troweling is applied after the concrete is hard enough so that mortar does not adhere to the edge of the trowel and a ringing sound is produced as the trowel passes over the surface. Additional trowelings may be desirable depending on the traffic and exposure conditions. During the final troweling, the trowel should be tilted slightly and heavy pressure should be exerted on the trowel to compact the surface thoroughly. Time should be allowed between successive trowelings to permit the concrete to increase its set.

Power troweling should be delayed until the concrete is firm. Power trowels are in wide use as finishing tools. As in steel troweling, premature power troweling must be avoided. The concrete must be hard enough to hold the power trowel.

PATTERNED AND TEXTURED FINISHES

A variety of patterns and textures are used to produce decorative finishes for concrete slabs. Patterns are formed by using divider strips or by grooving the surface just before the concrete hardens.

An exposed-aggregate finish provides a rugged nonskid surface. Selected aggregates, usually of a uniform 3/8-inch size or larger, are evenly distributed on the surface immediately after the use of the darby on the slab. The aggregate particles are embedded in the concrete by lightly tapping them with a darby or a flat board. After the concrete hardens sufficiently to support a finisher on kneeboards, the surface is hand floated with a magnesium float or darby. The aggregate is then exposed by simultaneously brushing and flushing the surface with water. For large jobs, a reliable retarder may be sprayed or brushed on the surface. On small jobs, however, this step may not be necessary. Since the timing is important in this process, test panels usually are made to determine the correct time for exposing the aggregate without dislodging the particles.

FINISHING AIR-ENTRAINED CONCRETE

Air-entrained concrete has a slightly altered consistency that requires some changes in the finishing operations as compared to those used with non-air-entrained concrete.

Air-entrained concrete contains microscopic air bubbles that tend to hold in suspension all of the materials in the concrete, including water. This type of concrete requires less mixing water than non-air-entrained concrete. However, air-entrained concrete still has good workability with the same slump. Since there is less water, and since it is held in suspension, there is little or no bleeding with air-entrained concrete. In the absence of bleeding, there is no need to wait for the evaporation of free water from the surface before the floating and troweling operations are started. This means that floating and troweling usually are started sooner for air-entrained concrete, before the surface becomes too dry or tacky. If floating is done by hand, the use of an aluminum or magnesium float is essential. A wood float drags and greatly increases the amount of work necessary to accomplish the same result. If floating is done by power, there is almost no difference between the finishing procedures for air-entrained and non-air-entrained concrete, except that floating can be started sooner on the air-entrained concrete.

Nearly all horizontal surface defects and failures are caused by the presence of bleed water or excessive surface moisture during the finishing operations. As a result, a better surface appearance is generally obtained with air-entrained concrete.

STUDY/DISCUSSION QUESTIONS

1. What is meant by finishing concrete?
2. Why is concrete consolidated? How is this done?
3. Explain why striking off is necessary. How is striking off performed?
4. Why should a small amount of concrete be kept ahead of the straightedge at all times?
5. What is the purpose of floating?
6. When should floating be started?
7. What are the tools used for floating?
8. When should edging and jointing be performed?
9. What are two other names for control joints?
10. Why are control joints necessary?
11. What is the recommended spacing for control joints in slabs? Sidewalks?
12. How deep are control joints cut?
13. If a saw is used, how soon can control joints be cut?
14. Explain the purpose of final finishing.
15. When should final finishing be started?
16. How can excessive surface water be removed from slabs?
17. What are some methods for obtaining a desired surface texture?
18. Why is it recommended that a steel trowel be held as flat as possible for the first troweling?
19. Why is a broken-in trowel used for the first troweling?
20. Why is air-entrained concrete preferred?
21. Why is it necessary to use an aluminum or magnesium float on air-entrained concrete?
22. Why are metal tools used only for low-slump concrete?

UNIT 15

CURING CONCRETE

Concrete hardens because of hydration, the chemical reaction between portland cement and water. As long as the temperatures are favorable and moisture is present to hydrate the cement, the following properties of concrete improve with age: durability (resistance to freezing and thawing), strength, watertightness, wear resistance, and volume stability.

Approximately three gallons of water are required to hydrate each bag of cement. Additional water is required, however, to place the freshly mixed concrete. Excessive evaporation of water from newly placed concrete can cause the cement hydration process to stop too soon. To prevent this loss of water, the concrete should be protected and curing started as early as possible. A rapid loss of water also causes the concrete to shrink. As a result, the tensile stresses are created at the drying surface. If these stresses develop before the concrete attains adequate strength, plastic shrinkage cracks may result.

EFFECT OF CURING

All of the desirable properties of concrete are improved by the proper curing process. Soon after the concrete is placed, the increase in strength is very rapid (for a period of 3 to 7 days). The strengthening then continues slowly for an indefinite period. Concrete which is moist cured for 7 days is about 50 percent stronger than that which is exposed to dry air for the same period. If the concrete is kept damp for one month, the strength is about double that of concrete cured in dry air.

METHODS OF CURING

Concrete can be kept moist by a number of curing methods. These methods can be divided into two classifications: those which supply additional moisture to the concrete, and those which prevent loss of moisture from the concrete by sealing the surface.

Water Curing

Curing by flooding, ponding, or mist spraying, is widely used. It is the most effective of all known curing methods for the prevention of mix water evaporation. This method is not always practical, however, because of job conditions. On flat surfaces such as pavements, sidewalks, and floors, the flooding or ponding method is easily accomplished. A small dam of earth or other water-retaining material is placed around the perimeter of the surface and the enclosed area is continuously flooded with water. Continuous sprinkling with water is also an excellent method of curing. If the sprinkling is done at intervals, the concrete must not be allowed to dry between applications of water. A constant supply of water prevents the possibility of crazing or cracking due to alternate wetting and drying.

Water Retaining Methods

These methods involve the use of coverings that are kept continuously wet, such as sand, burlap, canvas, or straw. When concrete is cured by one of these methods, the entire concrete surface, including exposed edges or sides, must be covered. For example, when the forms are removed, the sides of pavements and sidewalks must be covered. The materials used to retain the water must be kept damp during the curing period. If drying is permitted, excessive moisture may be absorbed by the cover itself. In vertically formed concrete, a simple way to prevent the concrete from drying out is to leave the forms in place.

Waterproof Mechanical Barriers

Barriers of waterproof paper or plastic film seal in the water and prevent evaporation. One important advantage of mechanical barriers is that periodic additions of water are not required. These materials are applied as soon as the concrete is hard enough to resist surface damage. The widest practical width of material is used. The edges of adjacent sheets are overlapped by several inches. The seams are then tightly sealed with sand, wood planks, pressure-sensitive tape, mastic, or glue.

An added feature of mechanical barriers is that they provide some protection against damage from subsequent construction activity as well as protection from the sun.

In some cases, plastic films may cause discoloration of hardened concrete. This is true for exterior or exposed concrete, and also when the concrete surface has been steel troweled to a hard finish. When discoloration is objectionable, some other curing methods must be used.

Chemical Membranes

Chemicals can be sprayed on the surface to cure concrete. Liquid membrane-forming curing compounds retard or prevent the evaporation of moisture from the concrete. These compounds are effective curing materials when used correctly. The chemical application should be made as soon as the concrete is finished. If there is any delay in the application, the concrete must be kept moist until the membrane is applied. The membrane curing compound must not be applied when there is free water on the surface. Free water will be absorbed by the concrete and the membrane will be broken. The compound should not be applied after the concrete is dry since it will be absorbed into the surface of the concrete and a continuous membrane will not be formed. The correct time to apply the membrane is when the water sheen disappears from the surface of the finished concrete. The adequate and uniform application of the curing compound is essential. In most cases, two applications are required. Chemical membranes are suitable not only for curing fresh concrete, but also for further curing the concrete after the removal of forms or after an initial moist curing.

There are four general types of curing compounds: clear or translucent, white pigmented, light gray pigmented, and black. During hot, sunny weather, white pigmented compounds are preferred since they reflect the sun's rays and thus reduce the concrete temperature. Curing compounds can be used to prevent the bond between hardened and fresh concrete. Consequently, these compounds are not to be used if a bond is necessary. For example, a curing compound should not be applied to the base slab of a two-course floor since it may prevent the top layer from bonding to the base layer. Some curing compounds also affect the adhesion of resilient flooring materials to concrete floors. Curing compound manufacturers should be consulted to determine if their products affect flooring.

GENERAL CURING REQUIREMENTS

In general, concrete should be cured for at least three days and preferably for a week after it is placed. The curing time depends on the temperature of the concrete. When the temperature is 70°F. or above, the concrete is cured for at least 3 days. When the concrete temperature is 50°F. or above, it is cured for at least 5 days.

The length of time that the concrete is to be protected against the loss of moisture depends upon several factors including the cement content, mix proportions, required strength, size and shape of the concrete mass, weather, and future exposure conditions. This protective period may be a month or longer for the lean concrete mixtures used in massive structures such as dams. This period also may be as short as a few days for richer mixes. Since all of the desirable properties of concrete are improved by curing, the curing period should be at least as long as is practical in all cases.

Concrete must be protected so that moisture is not lost too rapidly during the early hardening period. Concrete is kept at a temperature that is favorable for hydration. The most favorable temperature range for curing concrete is from 55° to 73°F. At higher temperatures, hydration takes place more rapidly, but the concrete does not attain its full strength. Hydration occurs at a slower rate when temperatures are below 70°F. There is almost no chemical reaction between cement and water when the temperature is near freezing. At 33° F., concrete requires more than three times as long to develop a given strength as it does at 70° F. No strength gain can be expected while concrete is frozen. When the concrete is thawed, hydration is resumed if suitable curing is applied. Freezing within the first 24 hours after the concrete is placed is almost certain to result in permanent damage to the concrete.

The concrete worker should remember that the longer concrete is moist cured, the better is its durability, strength, watertightness, wear resistance, and volume stability.

STUDY/DISCUSSION QUESTIONS

1. What causes concrete to harden?
2. Why must all concrete be cured?
3. About how many gallons of water are required to hydrate each bag of cement?
4. Why does concrete usually contain more water than is necessary for hydration?
5. Concrete which is moist cured for 7 days is about 50 percent stronger than concrete which is kept in dry air for the same period. If concrete is kept moist for 28 days, how does its strength compare with the strength of uncured concrete?
6. Explain some of the methods used for curing concrete.
7. Can discoloration result from using mechanical barriers for curing?
8. Is discoloration always objectionable?
9. Are there situations where chemical membranes should not be used for curing? Why?

10. What is the general length of time required to cure concrete when it is above 70°F.?
11. What is the general length of time required to cure concrete when it is above 50°F.?
12. What is the most favorable temperature range for curing concrete?
13. What will happen if concrete freezes during the first 24 hours after it is placed?
14. List five general characteristics of concrete that has been moist cured for a long period of time.

UNIT 16

JOINTS FOR FLAT CONCRETE WORK

Three basic types of joints are commonly used in slab-on-ground construction. These joints, which are called isolation, control, and construction joints, are easy to make and should be used in flat work such as floors, slabs, and walks.

WHY JOINTS ARE NECESSARY

All plastic concrete contains more water than is required for the hydration of the cement. When this extra water starts to evaporate, the process of drying-shrinkage of the slab creates tensile stresses in the concrete. These tensile stresses must be relieved by providing joints. If the stresses are not relieved, unsightly cracks will develop.

Concrete slabs expand and shrink by different amounts depending upon their shape and size. The larger the slab, the greater will be the horizontal movement. Provision must be made for this movement to prevent unnecessary cracking.

Slabs may have some vertical movement due to heaving or settling of the subgrade. The amount of movement may vary due to the lack of uniform subgrade bearing pressures, unequal loading conditions, changes in moisture content of the soil, and the alternate freezing and thawing of the subgrade.

Slabs may require joints that are not concerned with movement, but rather are used to ease the concrete placement. In many cases, this type of joint is combined with another type of joint.

ISOLATION JOINTS FOR POINTS OF ABUTMENT

Isolation joints are used to separate slabs from points of abutment such as walls, columns, and footings. This separation or isolation of the slab permits it to move without developing unsightly cracks.

To prevent bonding and provide a cushion, a layer of a continuous rigid asphalt-impregnated insulation strip, building paper, polyethylene, or similar material is placed upright between the edge of the slab and the immovable object (abutting construction). The joint material need not be more than 1/4 inch thick.

If watertight joints are desired, they can be formed with beveled wood siding or strips held away from the wall by wedges. The wedges and strips are removed after the concrete hardens. The space remaining is partially filled with sand and other resilient materials to provide a watertight joint.

CONTROL JOINTS TO PREDETERMINE CRACK LOCATIONS

Concrete tends to contract or shrink in cooler temperatures or when it loses moisture. Uncontrolled shrinkage in large slabs of concrete will cause random cracking at 15- to 25-foot intervals. The amount of cracking will depend upon such factors as the total amount of water in the concrete mix, the temperature and humidity ranges experienced, and the sub-base restraint. Therefore, it is good practice to predetermine the crack location. This is done by cutting the slab at selected locations to create a series of weakened planes. The weakened planes (joint locations) should be laid out so that the slab panels are approximately square. Slabs that are not rectangular or square should have a control joint across the slab at the juncture of the offset. Control joints generally are placed 10 to 15 feet apart on floor slabs, driveways, and livestock feeding floors. Control joints are placed 4 to 5 feet apart on sidewalks. Any cracks will follow the grooves or joints and thus random cracking is prevented.

The control joints are cut to a depth of 1/4 to 1/5 the thickness of the slab. The depth of cut is important. If the cut is too shallow, cracks will form elsewhere. If the cut is too deep or if it goes all the way through the slab, the property of load transfer may not be obtained. As a result, differential vertical movement will occur. With properly cut joints, aggregate interlock prevents unequal settling of the slabs. The cuts for the control joints should be made as soon as possible after the concrete is placed so that the larger pieces of aggregate can be worked away from the joint. A control joint can be cut simply by using a groover or jointer with a bit 3/4 to 1 inch deep. Another method of making a joint is to use a brick mason's trowel to cut the concrete. A groover is then used to finish the joint.

An electric- or gasoline-powered saw can be used to cut control joints. The joint is usually cut in the concrete surface 4 to 12 hours after the concrete hardens or as soon as the concrete surface will not be torn or damaged by the saw blade kerf.

CONSTRUCTION JOINTS FOR PLACING LARGE AREAS IN STAGES

A true construction joint should not be a plane of weakness nor an interruption in the mass of the concrete. Therefore, every effort must be made to insure the bonding of a new concrete area to the one placed previously. The old concrete should be cleaned thoroughly and dampened before the new concrete is placed to insure a bond. To avoid breaking the bond after the pavement is in use, tiebars or keyways frequently are used to assist in vertical transfer across the joint. Tiebars have the advantage of being able to resist shearing stresses caused by differential shrinkage of the old and new concrete.

COMBINATION CONTROL AND CONSTRUCTION JOINTS

A combination control joint and semiconstruction joint can be formed. The joint is constructed completely through the slab and is not just a weakened plane joint. As a result, the joint is a nonbonded and keyed control joint, figure 16-1. The bond between the new and previously placed slabs is broken by spraying or painting the concrete with a curing compound, asphalt emulsion, or form oil. This material allows the freedom of horizontal movement necessary to achieve crack control. The key provides the same type of vertical load transfer that is provided by the aggregate interlock in the cut joints. These transfer methods prevent unequal settling of the slabs.

Keyways can be made using a sheet metal form, premolded mastic 1/8 inch thick, composition hardboard, or 2-inch lumber with a metal strap or wooden strip attached to form the groove. If wooden strips are used to form the keyway, they may absorb moisture from the concrete, swell, and cause the concrete lip to crack above the keyway. To prevent cracking, the wooden strips should be well oiled and grooved with a saw kerf to minimize swelling. A 1 x 2-inch keyway with beveled edges is sufficient for slabs from 5 to 8 inches thick. Larger keyways should be avoided. If metal is used to form the keyway, it should be

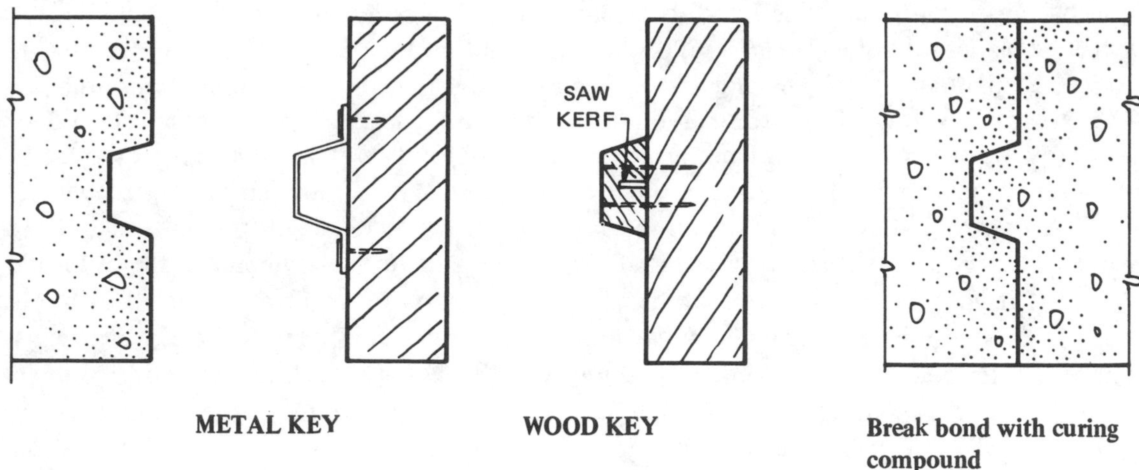

METAL KEY **WOOD KEY** **Break bond with curing compound**

Notes: A beveled 1" x 2" wood strip is adequate for most slabs.

Saw kerf groove prevents breakage of the concrete edge when moisture expands the 1" x 2" wood strip..

(Used when necessary to stop and join later.)

Keys of galvanized metal may be used.

Fig. 16-1 Combination control and construction joints.

thick enough to withstand handling. One advantage of the metal keyway form is the ease of cleaning for reuse.

STUDY/DISCUSSION QUESTIONS

1. Explain why joints are necessary for flat concrete work.
2. Where is an isolation joint used? Why is an isolation joint required?
3. How is an isolation joint formed?
4. How is a watertight joint obtained?
5. Why are control joints required?
6. What is the spacing for control joints?
7. How deep are control joints cut?
8. What occurs if the joint is too shallow? Too deep?
9. Explain the procedure for making control joints.
10. What is a construction joint and when is it used?
11. Where is a combination control-construction joint used?
12. When a wood strip is used to form the groove for a keyway, what precaution must be observed?
13. What size strip is adequate for 5- to 8-inch thick slabs?

UNIT 17

ESTIMATING CONCRETE FOR A JOB

The unit measure for concrete is the cubic yard, containing 27 cubic feet. Ready-mixed concrete is sold by the cubic yard. Concrete produced on the job is estimated by the cubic yard. To determine the amount of concrete needed for a job, the volume in cubic feet of the area to be concreted is first determined. This volume is then divided by 27 to obtain the value in cubic yards. The following formula can be used to determine the amount of concrete needed for any square or rectangular area.

$$\text{Cubic yards of concrete} = \frac{\text{width in feet x length in feet x thickness in feet}}{27}$$

For example, a 4-inch thick floor for a 30 x 60-foot building requires:

$$\frac{30 \times 60 \times 1/3}{27} = 22.22 \text{ cu. yd. of concrete}$$

The amount of concrete determined by the above formula does not allow for waste or slight variations in concrete thickness. Additional concrete in the amount of 5 to 10 percent must be added to cover waste and other unforeseen factors.

Note: In the above formula, the thickness dimension must be changed to feet or parts of a foot. For this example, 4 inches is changed to the fraction 1/3 foot. However, the decimal equivalent of the dimension may be used instead of the fraction. Table 17-1 gives both the fractional and decimal values for several common thickness dimensions.

Inches	Fractional Part of Foot	Decimal Part of Foot
4	4/12 or 1/3	0.33
5	5/12	0.42
6	6/12 or 1/2	0.50
7	7/12	0.58
8	8/12 or 2/3	0.67
10	10/12 or 5/6	0.83
12	1	1.00

Table 17-1 Thickness in inches expressed as a fractional or a decimal part of a foot for use in calculating quantities of concrete.

MATERIALS REQUIRED FOR JOB BATCHING

When concrete is job batched, the required quantities of materials (cement and fine and coarse aggregate) are estimated separately for each cubic yard of concrete needed. Table 17-2 shows the amount of portland cement in hundredweight, fine aggregate in cubic feet, and coarse aggregate in cubic feet required to produce 1 cubic yard or 27 cubic feet of mixed concrete for the suggested trial mixes. The example following the table shows how to compute the amounts of materials needed.

Suggested mixtures for 1 cwt. trial batches*			Materials per cu. yd. of concrete				
Cement	Aggregates		Cement	Aggregates			
cwt.**	Fine cu. ft.	Coarse cu. ft.	cwt.	Fine cu. ft.	Fine lb.	Coarse cu. ft.	Coarse lb.
With 3/4-inch maximum aggregate size							
1	2	2 1/4	7	17	1550	19.5	1950
With 1-inch maximum aggregate size							
1	2 1/4	3	5 1/2	15.5	1400	21	2100
With 1 1/2-inch maximum aggregate size (preferred mix)							
1	2 1/2	3 1/2	5 1/4	16.5	1500	23	2300
With 1 1/2-inch maximum aggregate size (alternate mix)							
1	3	4	4 1/2	16.5	1500	22	2200

*Mix proportions will vary slightly depending on the gradation of the aggregates. A 10-percent allowance for normal wastage is included in the above fine and coarse aggregate values.

**One cwt. of cement equals 1 cu. ft.

Table 17-2 Materials needed per cubic yard of concrete made with separated aggregates.

Example: How much material will be required to place a 4 inch thick concrete floor in a building 24 ft. x 14 ft.?

Find the number of cubic feet of concrete required. Multiply the length of the floor times the width of the floor times the thickness in feet. (4 in. = 1/3 ft.).

$$24 \times 14 \times 1/3 = \frac{112 \text{ cu. ft.}}{27} = 4.15 \text{ cu. yd.}$$

The shaded line in Table 17-2 shows that 1 cubic yard of concrete (1:2 1/4:3 mix) requires 5.5 hundredweight of portland cement, 15.5 cubic feet of fine aggregate, and 21 cubic feet of coarse aggregate. Therefore, 4.15 cubic yards of concrete will require 4.15 times these amounts.

4.15 x 550 = 2,282.5 lb. cement.

4.15 x 15.5 = 64 1/3 cu. ft. fine aggregate = 2 1/2 cu. yd.

4.15 x 21 = 87 1/4 cu. ft. coarse aggregate = 3 1/4 cu. yd.

SHORTCUTS FOR CONCRETE ESTIMATING

Depth, in.	Sq. ft.	Depth, in.	Sq. ft.	Depth, in.	Sq. ft.
1	324	4 3/4	68	8 1/2	38
1 1/4	259	5	65	8 3/4	37
1 1/2	216	5 1/4	62	9	36
1 3/4	185	5 1/2	59	9 1/4	35
2	162	5 3/4	56	9 1/2	34
2 1/4	144	6	54	9 3/4	33
2 1/2	130	6 1/4	52	10	32.5
2 3/4	118	6 1/2	50	10 1/4	31.5
3	108	6 3/4	48	10 1/2	31
3 1/4	100	7	46	10 3/4	30
3 1/2	93	7 1/4	45	11	29.5
3 3/4	86	7 1/2	43	11 1/4	29
4	81	7 3/4	42	11 1/2	28
4 1/4	76	8	40	11 3/4	27.5
4 1/2	72	8 1/4	39	12	27

Table 17-3 Concrete placed from a batch equal to 1 cubic yard

Concrete footing size	Lin. ft. of footing	Factor*
8 in. x 16 in.		x 3.5 =
10 in. x 20 in.		x 5.4 =
12 in. x 24 in.		x 7.7 =

*Includes adequate allowance for waste

Total ______________

Total divided by 100 = _______ cu. yd.

Table 17-4 Continuous footing.

Example: An 8-in. x 16-in. footing for a 20-ft. x 40-ft. building requires how many cubic yards of concrete?

The footing, in linear feet, equals 120 ft. This value is multiplied by 3.5 (factor from Table 17-4).

120 x 3.5 = 420 ÷ 100 = 4.20.

Use 4 1/4 cu. yd. of concrete.

Foundation thickness, in.	Height ft.		Length ft.		Factor*
6		x		x	2.0 =
8		x		x	2.6 =
10		x		x	3.2 =
12		x		x	3.8 =

*Includes adequate allowance for waste.

Total ______________

Total divided by 100 = __________ cu. yd.

Table 17-5 Continuous foundation.

Example: An 8-inch thick foundation, 5 feet high by 25 feet long, requires how much concrete?

5 ft. x 25 ft. = 125 sq. ft. (area)

125 x 2.6 (factor for 8 inch thickness from Table 17-5) = 325

325 ÷ 100 = 3 1/4 cu. yd.

or $\frac{5 \times 25 \times 2.6}{100}$ = 3 1/4 cu. yd.

Thickness of floor, in.	Area in sq. ft. (length x width)		Factor*
4		x	1.3 =
5		x	1.6 =
6		x	2.0 =

*Includes adequate allowance for waste.

Total ________________

Total divided by 100 = _______ cu. yd.

Table 17-6 Floor slabs.

Example: How much concrete is required for a 40 ft. x 60 ft. x 4 in. thick slab?

Refer to Table 17-6.

40 x 60 = 2,400 sq. ft. (area)

2,400 x 1.3 (factor from Table 17-6 for 4 inch thickness = 3,120

3,120 ÷ 100 = 31.2

Use 31 1/4 cu. yd.

USING A CONCRETE CALCULATOR

Small sliderules are used to make quick concrete calculations. In general, these sliderules are not accurately calibrated. However, they are very useful for checking estimates and for making quick estimates of small quantities of concrete.

STUDY/DISCUSSION QUESTIONS

1. What is the unit of measure for estimating concrete requirements?
2. What is the formula for estimating concrete requirements?
3. Is the amount of concrete determined by the formula adequate for all jobs? Why?
4. For job batching, which materials must be estimated separately?
5. How many cubic yards of concrete are required for the following?
   a. A grade beam 12 inches wide and 2 feet deep for a 35-ft. x 100-ft. building.
   b. A 6-inch floor slab for a 40-ft. x 135-ft. building.
   c. A 4-inch thick sidewall, 3 feet wide and 45 feet long.
6. To the amounts determined in question 5, what percentage is to be added for waste and slight variations in thickness?
7. Can the small sliderules used for concrete estimates be relied on for accurate estimates?
8. Of what value are concrete sliderules?

UNIT 18

REINFORCEMENT FOR CONCRETE

Reinforcement is the term used to describe the steel bars and small or large welded wire fabric positioned in concrete to increase its tensile strength.

Concrete has great strength in compression; that is, it can support great loads that are placed directly upon it. Steel bars or other metal reinforcement is required in concrete if it is to resist stresses or forces that tend to bend or pull it apart. The compressive strength of concrete is about 10 times greater than its tensile strength, figure 18-1.

When the metal reinforcement is used in concrete, the reinforcement withstands the tensile pull. The tensile strength can be made equal to or greater than the compressive strength depending on the amount of reinforcement used, figure 18-2.

KINDS OF REINFORCEMENT

Many materials have been tried as reinforcement in concrete. Steel is universally accepted and used. One important advantage of steel is that its contraction and expansion characteristics due to temperature changes are nearly the same as those of concrete.

Reinforcing steel can be purchased in the form of reinforcing bars or as welded wire fabric. Bars may be either smooth or deformed. Smooth bars usually have small diameters. Deformed bars have luglike ridges that increase the bond between the concrete and steel. The bars are available in standard sizes and are designated by a separate number for each size. The bar size selected for a job depends on the amount of tensile force the concrete will carry. Table 18-1 gives the standard bar sizes up to a diameter of 1 inch, the steel area that each size provides, and the weight of 100 feet of bar. Larger bars are available for extremely heavy construction.

Bars usually are supplied by the mill in 60-foot lengths. Local suppliers of building materials normally stock 20- or 40-foot lengths.

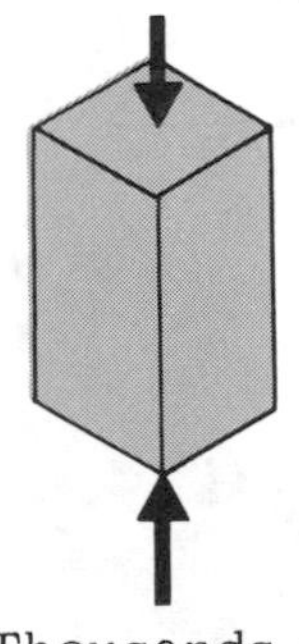

Thousands of pounds COMPRESSION

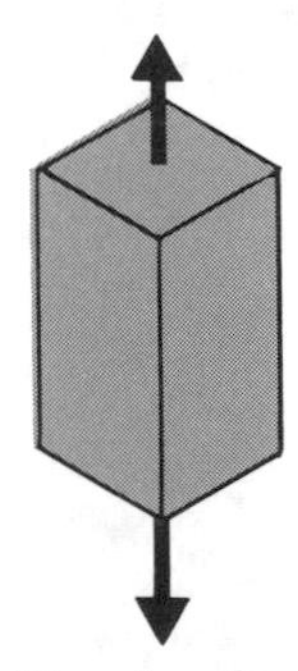

Hundreds of pounds TENSION

Fig. 18-1 Comparison of the compression and tension properties of nonreinforced concrete.

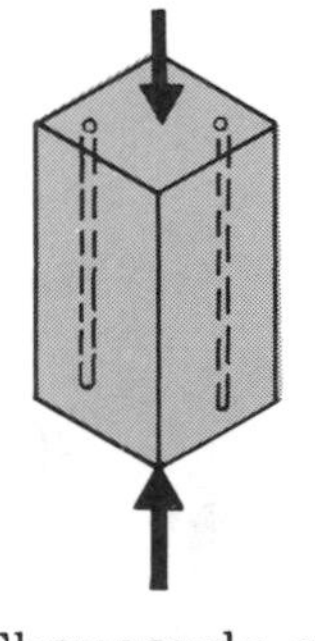

Thousands of pounds COMPRESSION

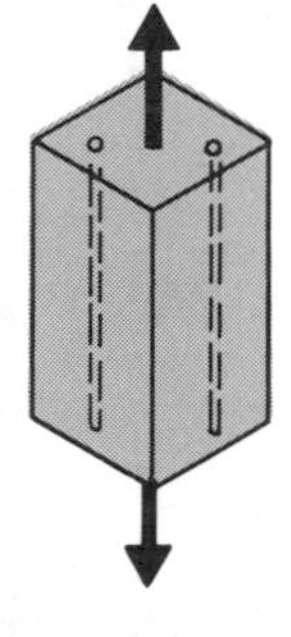

Thousands of pounds TENSION

Fig. 18-2 Comparison of the compression and tension properties of concrete when reinforcing is used.

Bar Number	Bar Diameter in.	Bar Area sq. in.	Approximate Weight of 100 Feet
2	1/4	0.05	17
3	3/8	0.11	38
4	1/2	0.20	67
5	5/8	0.31	104
6	3/4	0.44	150
7	7/8	0.60	204
8	1	0.79	267

Table 18-1 Size, area, and weight of reinforcing bars.

Fabric Style	Longitudinal Steel Area, sq. in. per ft.	Transverse Steel Area sq. in. per ft.	Weight lb. per 100 sq. ft.
6 x 6 No. 6	0.058	0.058	42
6 x 6 No. 8	0.041	0.041	30
6 x 6 No. 10	0.029	0.029	21

Table 18-2 Steel area and weight of three sizes of welded wire fabric.

Welded wire fabric is made in many types and sizes. The most common type of wire fabric used in light construction has wires spaced on 6-inch centers both horizontally and vertically. Wire fabric is commonly available in 6-, 8-, or 10-gage sizes. The type of fabric is usually written as 6 x 6 – 10 – 10 and is spoken as six-six-ten-ten. This expression indicates that the fabric has No. 10-gage longitudinal wires spaced on 6-inch centers, and No. 10-gage transverse wires spaced on 6-inch centers. Table 18-2 shows the area of steel provided by three sizes of wire fabric and the weight of 100 square feet of the fabric. Welded wire fabric is used for jobs that require relatively light reinforcement.

HOW REINFORCEMENT WORKS

Concrete can be subjected to a number of tensile forces including a straight tensile pull, bending, and forces resulting from temperature and moisture changes.

Concrete is often subjected to a *straight tensile pull* as in round structures such as water tanks and farm silos. Pressures within the tank tend to push the two halves apart, as shown in figure 18-3. Reinforcing steel in the wall holds the tank halves together. The steel must be added around the entire tank wall since the outward pressures act in all directions, not just in the directions shown in the figure.

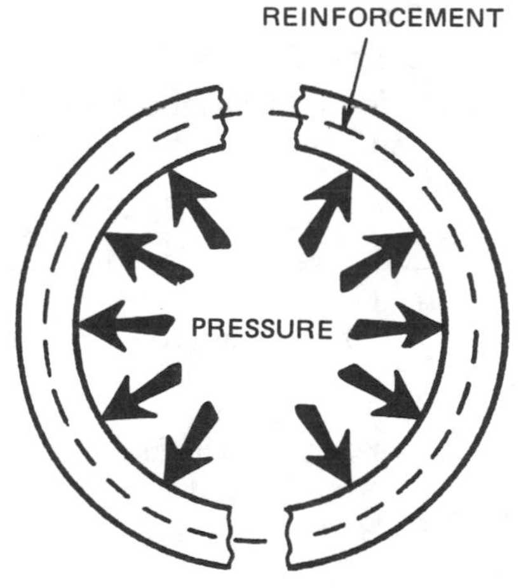

Fig. 18-3 Use of steel reinforcement to overcome the outward pressures within a round structure.

For round structures, the reinforcing steel is usually placed near the center of the wall cross

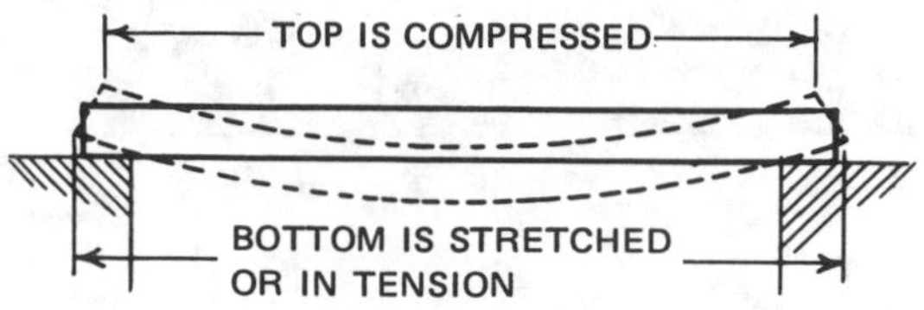

Fig. 18-4 Location of the compressive and tensional forces in a simple beam.

section, preferably slightly outside the centerline.

In concrete beams, only part of the beam contains tensile forces. Therefore, the location of the reinforcement is just as critical as the amount used to resist the bending loads. For example, in concrete lintels across door or window openings or in beams, the reinforcement is placed near the lower side. This is the side which tends to pull apart when the beam is loaded.

The lintel must be able to carry the load of the wall, floor, or roof that rests on it. The dotted lines in figure 18-4 show a greatly exaggerated picture of the bending that occurs in the lintel. Note that the ends of the beam at the top surface have moved closer together. The ends at the bottom surface have moved apart. Thus, it can be seen that the concrete at the top of the lintel is compressed and the concrete at the bottom is stretched or in tension. Since concrete cannot survive much tension, reinforcing steel must be placed near the bottom of the lintel.

TEMPERATURE AND MOISTURE CHANGES

Concrete expands and contracts with changes in temperature and moisture. Reinforcement can be used to reduce the cracking due to these changes. Welded wire fabric will not prevent the formation of cracks, but it will distribute them and generally make them smaller. For some concrete work, the slab may be jointed into 10- to 15-foot squares. This procedure is usually more economical than using reinforcement if the sole purpose of the reinforcement is to control temperature and shrinkage stresses.

PLACEMENT OF STEEL

The size, location, and spacing of the reinforcement are determined in advance by experienced engineers. The concrete worker on the job will perform the important operation of placing the reinforcement.

All reinforcement should be placed so that it is protected by an adequate coverage of concrete, Table 18-3.

Members	Minimum Concrete Protection
1. Footings	3 in.
2. Concrete surface exposed to weather	2 in. for bars larger than No. 5, 1 1/2 in. for No. 5 bars and smaller
3. Slabs and walls	3/4 in.
4. Beams and girders	1 1/2 in.
5. Joists	3/4 in.
6. Columns	Not less than 1 1/2 in. or 1 1/2 times the maximum aggregate size
7. Corrosive atmospheres or severe exposures	Protection shall be suitably increased

Table 18-3 Concrete protection for reinforcement*

*American Concrete Institute ACI 318, *Building Code Requirements for Reinforced Concrete*.

SPLICING REINFORCEMENT

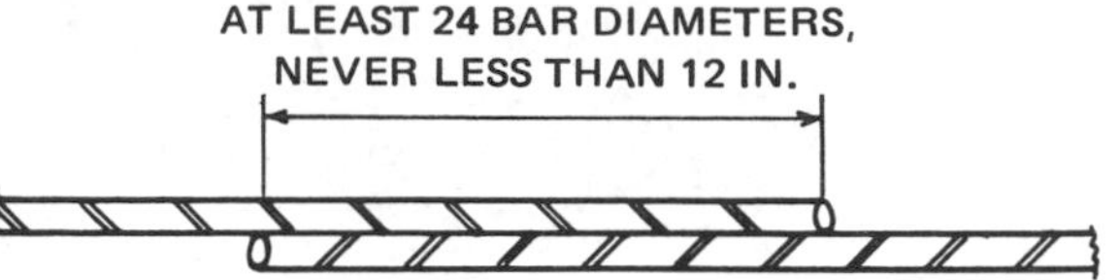

Fig. 18-5 Bars must be overlapped to transfer tensile stress.

Reinforcing steel carries a tensile load in concrete; therefore, the steel must be lapped at a splice. The overlap distance for deformed bars should always be at least 24 bar diameters, with a minimum lap of 12 inches, figure 18-5. Smooth bars must have an even greater overlap.

Welded wire fabric is spliced by lapping at least one full stay spacing plus 2 inches for a 6-in. x 6-in. fabric; this means a lap of at least 8 inches.

SLAB REINFORCEMENT

Reinforcement is used in floor slabs for homes under the following conditions:

1. If load-bearing partitions are more than 4 feet from the center axis of the slab.
2. If the slab is placed on fill more than 2 feet deep, or more than 10 percent of the area within the foundation wall is excavated and backfilled.
3. If heat ducts or pipes are embedded in the slab.
4. If unheated slabs are longer than 30 feet.

The use of reinforcing bars or welded wire fabric does not insure the prevention of cracks. However, the reinforcement may reduce the size of the opening if a crack does occur. The reinforcing steel should be placed 1 inch from the top of the slab and held in this position. If welded wire fabric is used, it should not be placed on the subbase and pulled up with a rake or hook. It is virtually impossible to control the position of the wire fabric by this method.

If the floor does not contain reinforcement but does have warm air ducts, the area over the ducts should be reinforced. A 6 x 6 x 10-gage fabric is used and is extended 18 inches past the point where the slab thickness returns to normal.

CONCRETE AROUND STEEL

Concrete should be placed around and under all reinforcement and embedded fixtures. The form can be tapped with a rubber or wood mallet (any similar practical method may be used) to make the concrete settle around and under the reinforcement.

GENERAL RULES FOR STEEL REINFORCEMENT

The following list gives the general rules to be followed in using steel reinforcement.

1. Use only clean steel free from rust, paint, and scale.
2. Place steel as recommended in Table 18-3, *Concrete Protection for Reinforcement.*
3. Limit the aggregate size used in the concrete to 3/4 the size of the minimum spacing between the reinforcing bars or between the reinforcement and the forms.
4. Lap all bars a minimum of 24 times their diameter and never less than 12 inches.
5. Place reinforcement where the concrete will be in tension.

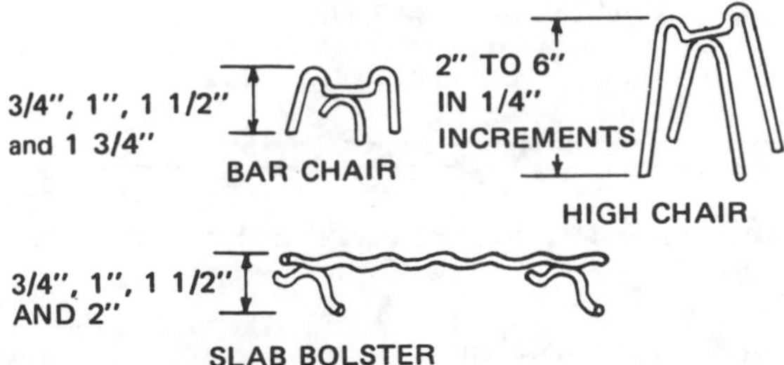

Fig. 18-6 Three of the more common devices available for easing the job of steel placement.

REINFORCED CONCRETE ACCESSORIES

Contractors' supply houses and concrete accessory manufacturers stock a wide range of materials to help ease the job of steel placement. Bar chairs, form ties, bar spacers, slab bolsters, and many other devices are available. Figure 18-6 shows some of the most common accessories.

NEW DEVELOPMENTS IN REINFORCEMENT

In recent years, great strides have been made in reinforced concrete technology. Reinforcing bars are constantly being improved. The deformed bars used today bond to the concrete much more firmly than those used only a few years ago. This improvement was achieved by changing the shape of the deformity or lugs. A recent development is high-strength steel reinforcing bars. When high-strength steel is used, the amount of reinforcing steel can be reduced.

STUDY/DISCUSSION QUESTIONS

1. What is reinforcement?
2. What is the difference between tension and compression?
3. Is concrete considered to be strong in compression?
4. Bar diameters are designated by number. What is the number of a 1/2-inch diameter bar?
5. What is the diameter of a No. 8 bar?
6. What are the three gages of fabrics commonly available?
7. When is welded wire fabric used?
8. Which side of a beam or lintel tends to pull apart when the concrete member is loaded, the top or the bottom?
9. Where is steel placed to counteract the load applied to a beam or lintel?
10. Most construction materials expand and contract with changes in temperature and moisture. Is this true of concrete?
11. Explain the rule for lapping bars.
12. Explain the rule for lapping wire fabric.
13. Explain how concrete can be worked around and under all steel reinforcement.
14. What determines the minimum distance steel is to be placed from an outside surface?
15. What is a bar chair?
16. What is a slab bolster?

UNIT 19

CONCRETING DURING HOT WEATHER

Spring and fall usually are considered the ideal seasons for concreting since these seasons do not have the temperature extremes of the winter and summer seasons. However, since construction continues year round, concrete must be handled and placed during hot weather. Concreting in hot weather poses some special problems but can be accomplished if certain precautions are observed.

PREVENTING RAPID EVAPORATION

For any summer concreting, the main precaution is to prevent rapid evaporation of water from the concrete. Thus, concrete must be protected during and after the placing and finishing operations so that the chemical reaction can proceed unrestricted. If too much water evaporates, then the hydration will not be complete.

Rapid drying may cause problems such as strength reduction and plastic shrinkage cracks. In extreme cases, a rapid loss of moisture from the concrete at the surface may cause cracks to form within the first day or so and often within the first few hours. In addition, the concrete may stiffen before it can be consolidated because of rapid setting of the cement and excessive absorption or evaporation of the mixing water. This condition makes it difficult to finish flat surfaces.

TEMPERATURE, RELATIVE HUMIDITY, AND WIND

Several factors affect the rate of evaporation of water from concrete. Evaporation is influenced by the concrete temperature, air temperature, relative humidity, and wind velocity. Even relatively small changes in these conditions may have a pronounced effect on the rate of evaporation, especially if the changes occur simultaneously.

For example, when the relative humidity changes from 90 to 50 percent (see Table 19-1, No. 2), the rate of evaporation is increased five times. If the humidity is reduced further to 10 percent, evaporation is increased almost nine times.

When both the concrete and air temperatures increase from 50 to 70°F. (Table 19-1, No. 3) evaporation is doubled. If these temperatures are increased further to 90°F., evaporation is increased four times.

With an air temperature of 40°F., the rate of evaporation is tripled when the concrete temperature is raised from 60 to 80°F. (Table 19-1, No. 5).

Wind velocity is also important. The rate of water evaporation is four times greater when the wind velocity increases from 0 to 10 mph, (Table 19-1, No. 1). When the wind velocity increases to 25 mph, evaporation is nine times greater.

The rate of evaporation is highest for the following conditions: when the relative humidity is low, when the concrete and air temperatures are high, when the concrete temperature is higher than the air temperature, and when the wind is blowing over the concrete surface. The combination of hot, dry weather and high winds (common in the summer months) removes moisture from the concrete surface faster than it can be replaced by normal bleeding.

CEMENT TEMPERATURE

The cement temperature has only a minor effect on the temperature of freshly mixed concrete. This is due to the low specific heat of cement and the relatively small amount of cement in the mix.

	Case No.	Concrete Temp., °F.	Air Temp., °F.	Relative Humidity %	Dew Point, °F.	Wind Speed	Drying Tendency lb./sq.ft.,/hr.
(1) Increase in Wind Speed	1	70	70	70	59	0	.015
	2	70	70	70	59	5	.038
	3	70	70	70	59	10	.062
	4	70	70	70	59	15	.085
	5	70	70	70	59	20	.110
	6	70	70	70	59	25	.135
(2) Decrease in Relative Humidity	7	70	70	90	67	10	.020
	8	70	70	70	59	10	.062
	9	70	70	50	50	10	.100
	10	70	70	30	37	10	.135
	11	70	70	10	13	10	.175
(3) Increase in Concrete and Air Temperatures	12	50	50	70	41	10	.026
	13	60	60	70	50	10	.043
	14	70	70	70	59	10	.062
	15	80	80	70	70	10	.077
	16	90	90	70	79	10	.110
	17	100	100	70	88	10	.180
(4) Concrete at 70°F. Decrease in Air Temperature	18	70	80	70	70	10	.000
	19	70	70	70	59	10	.062
	20	70	50	70	41	10	.125
	21	70	30	70	21	10	.165
(5) Concrete at High Temp.; Air at 40°F. and 100% R. H.	22	80	40	100	40	10	.205
	23	70	40	100	40	10	.130
	24	60	40	100	40	10	.075
(6) Concrete at High Temp.; Air at 40°F. Variable Wind	25	70	40	50	23	0	.035
	26	70	40	50	23	10	.162
	27	70	40	50	23	25	.357
(7) Decrease in Concrete Temp.; Air at 70°F.	28	80	70	50	50	10	.175
	29	70	70	50	50	10	.100
	30	60	70	50	50	10	.045
(8) Concrete and Air at High Temp.; 10% R.H.; Variable Wind	31	90	90	10	26	0	.070
	32	90	90	10	26	10	.336
	33	90	90	10	26	25	.740

Table 19-1 Effect of variations in concrete temperature, air temperature, relative humidity, and wind speed on drying tendency of air at job site.

The cement loses heat slowly during storage. As a result, it may be warm when delivered. This heat is produced during the grinding of the cement clinker in manufacture. Since the temperature of cement does affect the temperature of fresh concrete to some extent, some specifications place a limit on its temperature at the time of use. However, test results indicate that it is more desirable to specify a maximum temperature for freshly mixed concrete than to place a limit on the temperature of the ingredients.

POINTS TO REMEMBER BEFORE PLACING AND FINISHING CONCRETE

A few simple precautions will insure a quality concrete job, and may save many dollars. The following procedures will give the concrete an earlier strength gain, greater durability, and will greatly reduce the occurrence of surface defects.

- Plan ahead. Before beginning extensive concreting operations, check with the weather bureau to obtain the latest forecasts for temperature, wind velocity, and relative humidity. Remember that these factors, singly or in combination, will influence excessive evaporation of water from the concrete surface.
- Be prepared with the necessary equipment and material well in advance of hot weather.
- Be sure of an ample water supply for use in sprinkling the subgrades, wood forms, and aggregates, and for curing.
- Have tarpaulins or polyethylene sheets and lumber ready for sunshades and windbreaks.
- Schedule work so that the concrete can be placed with the least delay. See figure 19–1. Jobs started late in the afternoon during extremely hot periods can take advantage of lower evening temperatures and decreased winds.

Fig. 19-1 Schedule work. (*Photo by Richard T. Kreh Sr.*)

- Use cool materials if the materials are to be mixed on the job, keep all materials cool by storing aggregates in the shade if possible, sprinkling coarse aggregate with water, and protecting the water supply from the direct sun rays. The mixing water may be chilled in very hot weather by refrigeration or by using ice as a part of the mixing water. The ice should be melted by the time the concrete leaves the mixer. Most ready-mixed producers follow these procedures so that cool concrete is delivered to the job site.
- Prevent absorption. Thoroughly moisten the subgrade, reinforcing steel, and wood forms just before the concrete is placed so that these items will not absorb water from the mix.
- Sprinkling cools the surrounding air and increases its relative humidity. This not only reduces the temperature rise of the concrete but also minimizes the evaporation of water during the placing operation. For slabs on ground, it is a good practice to dampen the subgrade the evening before concreting. The subgrade should be checked prior to the placing of the concrete to insure that there is no standing water or puddles on the subgrade.
- Coarse aggregates should be sprinkled before they are added to the batch.

POINTS TO REMEMBER DURING AND AFTER PLACING THE CONCRETE

- After the concrete is placed, strike it off and darby or bull float it at once.
- Place temporary covers, such as burlap kept continuously wet, over the fresh concrete immediately after striking and darbying.
- When ready for floating and/or final finish, uncover only a small section of the concrete immediately ahead of the finishers. Recover the concrete right after the final finish and keep the cover wet.
- Any delays in finishing air-entrained concrete in hot weather usually lead to the formation of a rubberlike surface which is difficult to finish without leaving ripples or ridges.
- Protect the concrete surface against evaporation. When placing concrete under a hot sun and in drying winds, the rapid loss of water must be prevented. A severe water loss may cause plastic shrinkage cracks.
- Protect concrete by placing a windbreak on the windward side when high winds prevail. It may be necessary to fog the air over new concrete with a mist of water.
- Protect the concrete from direct sunlight on hot days. Erect a sunshade or delay the placing operations until late in the day. Where possible, take advantage of the shade from nearby buildings or trees.
- Start curing as soon as the concrete surface is hard enough to resist marring. Cover the concrete with polyethylene film, waterproof paper, waterholding materials such as straw or burlap, or spray a curing compound on the concrete. If a curing compound is to be used, apply it immediately after the final finishing. Adequate and uniform coverage must be obtained.
- Constantly wet the concrete surface to avoid alternate wetting and drying during the curing period. See figure 19–2.

Fig. 19-2 Keep concrete surface wet. (*Photo by Richard T. Kreh Sr.*)

- **Continue curing for at least 3 days and preferably for a week. Water not only acts as a curing agent but also cools the slab.**

RECORDING WEATHER CONDITIONS

The weather conditions during the concreting should be recorded so that they are a part of the permanent job record. The humidity, temperature, wind, and clouds should be noted.

TESTING SPECIMENS IN HOT WEATHER

In hot weather, sampling (making and curing test specimens) must be done in conformance with the standard specifications. ASTM Designation C31, *Practice for Making and Curing Concrete Test Specimens in the Field,* governs the procedure to be used for specimens sampled from concrete being used in construction.

Test cylinders must be kept shaded and damp. When the test cylinders are one day old, they must be transferred to the laboratory (or other suitable location) where they will receive continuous standard moist curing until they are tested. A damp sand or wet burlap cover, or fog sprays will insure retention of the water by the specimens.

USE OF ADMIXTURES

Admixtures are sometimes used during hot weather to delay the setting time of concrete and lessen the need for an increase in mixing water. Water-reducing agents may be helpful if they do not interfere with strength development and other properties of the concrete. The use of water-reducing agents must be carefully controlled. These substances should be used to supplement, but not replace, other hot weather concreting procedures.

If admixtures are used, they should meet ASTM or federal specifications. Admixtures must be tested with job-site materials under job conditions, including temperature. The testing must be done in advance of construction to determine their compatibility with the other materials and their ability under these conditions to produce the desired properties.

STUDY/DISCUSSION QUESTIONS

1. What seasons are usually considered to be ideal for concreting? Why?
2. What adverse effects do temperature, relative humidity, and wind have on concrete during hot weather?
3. What may happen to concrete in hot weather if precautions are not taken?
4. What planning is necessary before placing and finishing concrete in hot weather?
5. What is meant by "use cool materials"?
6. What can be done to prevent water from being absorbed from the mix?
7. What operations should be completed as soon as possible after concrete is in place?
8. What can be done to protect the concrete surface against evaporation?
9. Is curing necessary during hot weather? Why?
10. How long should concrete be cured during hot weather?
11. What procedures should be followed when securing test specimens during hot weather?
12. If admixtures are used in concrete, what specifications or standards are recommended?

UNIT 20

CONCRETING IN COLD WEATHER

Concrete can be placed during the winter months if certain precautions are taken. Adequate protection must be provided for the concrete when temperatures of 40°F. or lower occur during placing and the early curing period.

Plans should be made well in advance to protect fresh concrete from freezing and maintain the minimum permissible curing temperatures. Appropriate equipment should be at hand and ready to use for heating the concrete materials, for constructing enclosures, and for maintaining favorable temperatures after the concrete is placed. The cost of construction will be increased somewhat by these essential precautions and will vary with the amount of protection required.

EFFECT OF CONCRETE TEMPERATURES

Temperature has a considerable effect on the rate of hardening of concrete. Temperature also affects the rate at which the hydration of cement occurs. Low temperatures retard concrete hardening and strength gain. Near the freezing point, the strength gain rate is very slow, and at temperatures below freezing there is almost no increase in strength. At temperatures below 73°F., strengths are lower at early ages but higher at later periods. Concrete made and cured at 55°F. has relatively low strength for the first few days. After 28 days, however, the concrete has slightly higher strength than concrete made and cured at 73°F. Concrete made at 40°F. and cured for 28 days at 25°F. has very little strength at early ages. However, if favorable curing is then provided, this concrete develops strength comparable to that of concrete cured at 73°F.

Strength gain almost stops when the moisture required for curing is no longer available. Concrete placed at low temperatures (above freezing) may develop higher strengths than concrete placed at high temperatures. However, the curing of concrete at low temperatures must continue for a longer period.

HOW TO OBTAIN HIGH-EARLY-STRENGTH CONCRETE

At times it is desirable to increase the setting rate and strength development of concrete. High strength at an early age is frequently desirable during winter construction to reduce the length of time that protection is required. High-early-strength concrete may be obtained by using one or a combination of the following:

1. High-early-strength cement
2. Additional portland cement
3. Higher curing temperatures (that is, steam curing).

The use of high-early-strength cement during cold weather is economical since it can mean earlier reuse of the forms and result in savings in the cost of additional heating and protection, and earlier use of the finished concrete.

USE OF CHEMICAL ACCELERATORS

Chemical admixtures known as accelerators may be used to shorten the setting time and increase early strength development of concrete mixes.

Calcium chloride is the most commonly used accelerating admixture. If used in concrete, it should be added in solution as part of the mixing water. However, the solution should not come into direct contact with the cement or a flash set may occur. If calcium chloride is added to the mix in dry form, undissolved lumps may cause popouts or dark spots in the hardened concrete. When ready-mixed concrete trucks must travel a long distance, the addition of the calcium chloride solution is sometimes delayed until after the concrete is mixed, shortly before discharge. Twenty revolutions of the drum are usually enough to incorporate the calcium chloride thoroughly in the mix.

The American Concrete Institute recommends the use of 1 percent calcium chloride by weight of cement for cold weather concreting when the mean temperatures are below 40°F. Calcium chloride in amounts up to 2 percent by weight of cement is sometimes used. However, calcium chloride in amounts greater than 2 percent can cause problems such as flash set, an increase in drying shrinkage, and reinforcement corrosion. If the concrete is watertight, the addition of calcium chloride in recommended amounts has no significant corrosive effect on ordinary steel reinforcement.

Calcium chloride or admixtures containing soluble chlorides must not be used in the following situations.

1. In concrete for prestressed concrete construction. Corrosion of the prestressing strands may result.
2. In concrete containing embedded aluminum, such as conduit. Serious corrosion may result, especially if the aluminum is in contact with embedded steel and the concrete is in a humid environment.
3. In lightweight insulating concrete placed over metal decks.
4. In concrete that will be in contact with soils or water containing sulfates.

DON'T DEPEND ON CHEMICALS TO PREVENT FREEZING

After concrete is in place it should be protected against freezing. However, do not use antifreeze compounds or other materials to lower the freezing point of the concrete. Such large quantities of these materials are needed to lower the freezing point of concrete appreciably that strength and other concrete properties are seriously affected. There is no such thing as an antifreeze for concrete. The concrete worker is cautioned not to depend on chemicals to prevent freezing.

When several freezing and thawing cycles occur at an early age, the strength and other qualities of the concrete are permanently affected. However, the strength of concrete that is subjected to a light freezing at an early age may be restored to normal by resuming the favorable curing conditions. This concrete, however, will not have the resistance to weathering, nor will it be as watertight, as concrete that is not frozen.

POINTS TO REMEMBER BEFORE PLACING AND FINISHING CONCRETE IN COLD WEATHER

- Plan in advance. Have equipment and materials ready before cold weather arrives. Provide heaters, insulating materials, and enclosures. Use high-early-strength concrete where job conditions make it desirable.
- Heat the materials. The temperature of the concrete as it is placed in the forms should range between 50° and 70° F. for slabs. When the air temperature is between 30° and 40° F., the mixing water should be heated. When the air temperatures are below 30° F., the mixing water and sand (and sometimes the coarse aggregate) should be heated. There should be no frozen aggregate lumps in the concrete when it is placed.
- To prevent flash set, materials should not be overheated. The maximum allowable water temperature is about 140° F.
- Do not place concrete on frozen ground since unequal settling will occur when the ground thaws. This can cause cracking.
- Fresh excavations should be covered with straw or other insulating material to prevent the ground from freezing until the concrete can be placed.
- Remove all ice and frost from forms and steel reinforcing.
- Use accelerators carefully. Use about 1 lb. of calcium chloride per bag of cement to hasten hardening. No more than 2 lb. of accelerator should be used because of the danger of flash set.
- Do not use any admixtures to prevent the concrete from freezing.
- Do not use calcium chloride with other admixtures which accelerate hardening.

POINTS TO REMEMBER AFTER THE CONCRETE IS PLACED

- Provide suitable curing temperatures. When using normal portland cement, the temperature of the concrete is to be maintained at 70° F. or above for 3 days, or 50° F. or above for 5 days.

The temperature of high-early-strength concrete is to be maintained at 70°F. or above for 2 days, or 50° F. or above for 3 days.

Do not allow the concrete to freeze during the next 4 days.

Cool the concrete gradually at a rate of 1 to 2 degrees per hour until it reaches the outside temperature.

- Keep job condition records. Record the date, hour, weather conditions, and temperature (both of the air surrounding the concrete and the surface of the concrete) at least twice daily.
- Protect the concrete. Insulation, such as a thick blanket of straw but no artificial heat, is often sufficient protection for slabs on ground. At lower temperatures, a housing and artificial heat are necessary.

Housings can be made of wood, insulation board, waterproof paper, or tarpaulins over wood frames.

Circulate moist warm air between floor slabs and the housing.

Avoid the risk of fire by placing coke- or oil-fired heating units away from flammable material. Vent the heating units to the outside.

Raise heating units above the floor to avoid rapid drying of the concrete underneath.

Keep the concrete moist, especially near heating units. In cold weather, when artificial heat is applied, moisture for curing is still very important. First wet the slab well with water and then cover the slab with waterproof paper. Apply heat to keep the slab from freezing. This water treatment plus the covering prevents the surface of the slab from drying out.

CURING METHODS

Concrete placed in forms or covered with insulation seldom loses enough moisture at 40° to 55° F. to impair curing. However, moist curing is needed during winter concreting to offset the drying tendency when heated enclosures are used. It is important that concrete be supplied with enough moisture when warm air is used to maintain favorable temperatures.

Live steam exhausted in an enclosure is an excellent method of curing because it provides both heat and moisture. Steam is especially practical during extremely cold weather because moisture provided by the steam offsets the rapid drying that occurs when very cold air is heated.

Early curing using liquid membrane-forming compounds is permitted on concrete surfaces within heated enclosures. It is a better practice, however, to cure the concrete first with water or exhaust steam during the period when it is first protected. Then a curing compound can be applied after the protection is removed and the air temperature is above freezing.

Heat may be kept in the concrete by using commercial insulating blankets or bat insulation. The effectiveness of the insulation can be determined by placing a thermometer in contact with the concrete under the insulation. If the indicated temperature falls below the minimum required temperature, additional insulating material must be applied. The corners and edges of concrete are the most vulnerable to freezing and should be checked to determine the effectiveness of the protection.

Recommendations on the amount of insulation necessary to protect concrete at various temperatures may be obtained from the manufacturers of these materials. The American Concrete Institute specification, *Recommended Practice for Winter Concreting* (ACI 604), also gives recommendations for insulation.

Heated enclosures are commonly used to protect concrete when the air temperatures are near or below freezing. There are a number of suitable materials for making enclosures.

Enclosures may be heated by live steam, steam in pipes, oil-fired burners, salamanders, and other types of heaters. Salamanders are easily handled and are inexpensive to operate. They are convenient for small jobs but have several disadvantages. Salamanders produce a dry heat, so care must be taken to prevent drying of the concrete, especially near the heating element. When placed on floor slabs, salamanders should be elevated and the concrete near them protected with damp sand.

Salamanders and other fuel-burning heaters produce carbon dioxide. This compound combines with the calcium hydroxide in fresh concrete to form a weak layer of calcium carbonate. When this occurs, the surface of the concrete floor will dust under traffic. For this reason, carbon dioxide-producing heaters should not be used while placing concrete and for the first 24 to 36 hours of the curing period, unless the heaters are properly vented.

REMOVE FORMS AT PROPER TIME

During cold weather, ample time must be allowed for the concrete to attain its required strength before the forms are removed. If the forms are removed too soon, the corners and edges of the concrete may chip. Forms should remain in place until the concrete attains sufficient strength to sustain its own weight in addition to any other load that may be placed upon it during construction.

DEFECTS: THE CAUSES AND CURES IN CONCRETE

DEFECT	CAUSE	CURE
Segregation—separation of coarse and fine aggregates.	Excessive mixing water. Use of tampers or 'jitterbugs.' Using darbies or floats too quickly on surface.	Use lower slump mix as 4-in. or less; use richer mix (more cement); use air-entrainment; don't tamp or jitterbug fresh concrete.
"Pick-up"—fresh concrete that sticks or clings to float, trowel, or knee-boards.	Segregation of sand and cement will cause sticking.	Try refloating the surface so sand and cement are re-mixed and evenly distributed.
Crazing—fine hair-line cracks in the surface of the hardened concrete. Sometimes called 'alligator' or map cracks. (Such concrete is usually not structurally unsound, although poor in appearance.)	Surface shrinkage resulting from rapid or excessive floating or troweling. Rapid drying of concrete surface due to a combination of high temperature, low humidity, and wind. Use of dry cement sand mixture as surface topping. Too high slump, excessive bleed water, premature troweling. Poor curing practices as wet/dry curing cycles.	Dampen subgrade prior to concreting. Provide protection during finishing, use good finishing techniques and proper curing. Use fog spray or cover with continuously wet burlap. Start curing immediately. Delay all finishing operations until all 'free' or 'bleed' water has left the surface.
Scaling—when concrete surface peels off or chips away in thin layers.	Repeated cycles of freezing and thawing in non air-entrained concrete. Use of de-icing salts. Over finishing. Faulty workmanship such as darbying, floating, and finishing while bleed water is on surface. Porous aggregates. Wet mix.	Use air-entrained concrete—6% is optimum. Use adequate compressive strength mix—a 6-bag cement per cu. yd. mix or 4,000-lb. psi. Place concrete at temperatures above 55°F. with a 4-in slump. Finish and cure properly.
Blisters—bubbles or areas in the concrete surface that appear as bumps or irregular raised areas appearing in thin layers or bulges.	Entrapment of air or bleed water under an air tight or sealed surface due to overfinishing, no vibration or early finishing.	Blisters may occur during the second or third pass in the finishing operation as air may be forced ahead of the trowel blade and under the skin of the concrete (cement paste). Delay troweling. Floating concrete the second time may help prevent blisters.
Plastic Shrinkage—cracks that occur when rate of evaporation is greater than the movement of water to the surface from within the concrete.	Rapid evaporation of water from the surface of the concrete caused by high temperatures combined with low humidity and wind movement. The internal temperature of the concrete may be too high. Don't exceed 75°F.—use a thermometer.	Dampen subgrade, protect concrete when pouring from wind and sun. Cure properly. Shorten up placing and finishing time. Unload trucks quickly. Use fog spray, plastic, or wet burlap between placing and finishing times.

continued

DEFECT	CAUSE	CURE
Discoloration—color changes in hardened concrete. Spotted or dark and light blotches on surface. (See also stains.)	Variations in slump. Sprinkling of dry cement on wet concrete surface. Too much steel troweling. Improper curing. Uneven curing as wet and dry burlap. Use of different concrete making materials in various batches for same job. Use of calcium chloride (accelerator for setting) where chloride reacts slowly with the ferrite ions thus leaving dark patches. Laying plastic or mats directly on concrete may discolor. Delayed hard troweling by machine may give black, dense surface.	Flush hardened concrete with water and wash with 3% phosphoric acid. Use proper slump. Do not use dry cement on fresh slab. Trowel at right time and with proper blade setting. Overlap plastic sheeting and seal edges. Apply spray-on curing compound in two-directions. Keep plastic or mats away from surface if color is important. Use same cement source for whole job. Use same aggregate source for whole job. Don't overtrowel with power trowel, leaving some areas more dense than others. Try paint or floor wax to hide defect.
Slick Concrete—hardened concrete that's slippery especially when wet or covered with damp leaves.	Over-finishing of concrete surface. Hard or machine troweling of surface, especially that exposed to outside elements.	Acid-etch surface to give desired roughness or texture. Sandblast or lightly saw surface. Use bushammer or scrabbler to roughen; surface. Use epoxy/sand topping following manufacturer's suggestions.
Honeycomb—rock pockets and voids in hardened concrete. Generally observed after removal of forms.	Poor mixing of concrete. Poor gradation of mix. No vibration or poor consolidation.	Use proper vibration or spudding techniques. Tap side-forms to work out air pockets and bring paste to surface. Patching may be necessary. Wet area and use rich cement/sand/water mix—use dry and tamp in.
Bleeding—excess or watery surface after strike-off of fresh concrete.	Soupy mix (too much water in mix). Poor or badly graded aggregate. Use of non-air entrained concrete. Overworking surface too early. Plastic sheeting under concrete.	Use 4-in. or less slump. Must have properly graded, well mixed aggregate. Use 6% air entrained concrete. Consider sand bed over plastic under concrete.
Slow To Set—failure of concrete to set-up or harden.	Cold sub-grade, forms or cold concrete.	Order calcium chloride in mix as accelerator...(Note manufacturers restrictions on its use.) Heat water and aggregates in mix. Heat building. Protect subgrade from freezing with insulated blankets.
Scaling—when concrete surface peels off or chips away in thin layers.	SCALING happens when there are repeated cycles of freezing and thawing in concrete that has not been air-entrained.	Use new cure-seal compounds and spray or roll-on at 200-sq. ft. per gallon. Do not overwork surface. Delay finishing until all bleed water has left surface. Do not sprinkle raw cement powder on fresh concrete to absorb bleed water. Prevent freezing of fresh concrete with insulated blankets. Don't add water to the mix at the jobsite. Don't apply salt to newly placed concrete the first year—use sand for traction. Never use ammonimum nitrate or ammonimum sulfate as de-icing salts. In severe cases remove topping mechanically and cure seal or apply epoxy topping according to manufacturer's recommendations.
Dusting—a powdery, chalky material that appears on the dry surface of hardened concrete. Soft, easily scratched surface is eroded under traffic.	Excessive slump (high water to cement ratio). Finishing while bleed water is on surface. Inadequate curing. Premature finishing brings up excess bleed water to surface	Use 4-in. or less slump concrete. Finish when 'bleed-water' has evaporated from the surface. Vent or exhaust heaters and engines. Use concrete above 55°F. when problems are expected. If possible

DEFECT	CAUSE	CURE
	and thins out paste. Carbonation due to burning salamanders, heaters or running engines in enclosed area. Condensation of warm moist air over cool concrete. Not sufficient moisture to allow cement in mix to hydrate properly. Over abundance of 'fines' in the sand gradation. Low-strength concrete.	flood area with warm water and cure wet to repair. Use floor hardener as zinc or magnesium flurosilicate as directed by manufacturer. Wet grinding of surface is another possible repair.
Cracks—breaks or separation in the concrete slab other than joints. Concrete that pulls apart leaving separations or cracks. (Also, see crazing and shrinkage.)	Contraction (shrinkage) of concrete due to drying or cold weather. Movement of subgrade. Poor subgrade preparation or uneven subgrade settlement. Volume change in the concrete. Hydrostatic or water pressure. Heavy loads applied to slab. Cracks that occur before concrete hardens mostly due to settlement of the mass or shrinkage. After concrete hardens, cracks result from drying shrinkage, contraction, or loading.	Provide isolation joints by separating fresh concrete from hardened concrete with material. Isolate poles, posts, drains, and the like. Cut control joints closely to provide controlled cracking. Cut appx. 1/4 slab thickness not more than 15-ft. apart on slabs. Use proper water/cement ratio for job (rich mix). Use low-slump concrete and cure. Compact subgrade. Don't place on frozen subgrade. Provide adequate drainage.
Spalling—surface fragments sloughing off. Flaking of concrete.	Generally caused by a mechanical force on concrete surface—such as a blow. Action of weather, expansion. Rusting of re-bars in slab or restraint of slab.	Cut proper joints. Use keyed joints or joints properly cut and prepared. Consider sawed joints spaced closer together. Have minimum of 2-in. of concrete over reinforcing bars.
Popout—small conical areas that break out of the concrete surface leaving shallow depressions.	The expansion of aggregate used in the mix materials. Soft aggregate absorbs water and expands until it pops out leaving a hole or pockmark. Water absorbed aggregate may freeze, expand and popout.	Use mix with manufactured aggregates if natural aggregates have excessive reactive material. Watch mix for flat, poorly-shaped aggregates. If severe aggregate problem, consider 2-course construction with poorer material on bottom topped with a higher quality mix.
Efflorescence—a deposit of white, powdery material on surface of hardened concrete.	Migration of material, usually salts, from below the surface that forms when the water-carrier evaporates, thereby leaving deposits.	Most effective treatment is prolonged washing, flushing, and scrubbing with water. Scrub with non-metal brush or broom. Try flushing with a dilute 3% solution of acetic acid or 2% phosphoric acid following manufacturer's recommendations for application and safety.
Form Failure—most often found in vertical forming systems where concrete is not contained to conform to required shape or size.	Inadequate forming materials, form ties, installation or bracing. Filling forms too rapidly. Careless vibration.	Select manufactured forms to meet design use. Use properly spaced and adequate form ties—don't skip recommended spacing. Use adequate wallers and proper external bracing. Don't fill forms too rapidly. Start and place concrete in 'lifts.' Vibrate or spud concrete but watch for form ties and re-bars.
Blow-Up—rapid setting of concrete surface. Premature hardening.	Rich concrete mix. Hot, dry, windy days. Dry sub-base. Hot concrete mix or one mixed too long in truck. Delay in placing mix.	If this problem occurs consistently, consider ordering retarder in mix. Have crew ready to act in emergency. Get subgrade pre-wetted and keep damp. Erect windbreak or sunscreen. Screed, float and finish as required—leaving joints to be

continued

DEFECT	CAUSE	CURE
		saw-cut later. Use the fog cure or cover it up between screeding/floating/finishing operations.
Rain–rain hitting fresh concrete may cause surface damage.	Natural occurence.	Check on weather conditions prior to scheduling pour. Have protection materials and equipment on hand at the jobsite. Cover and protect concrete at end of day. Have rolls of plastic for protection. Straw may be used in emergency but tends to stain. After sudden, heavy shower use rubber garden hose to "draf-off" excess rain water before final finishing.
Freezing–Fresh concrete subjected to low temperatures.	Natural occurence. Poor job planning. Putting concrete on frozen subgrade or in frozen, ice-covered forms. No protection.	Concrete can be placed in freezing weather–even below 0°F. if the subgrade is not frozen and the materials can be protected from freezing by insulated blankets or enclosures that are heated. Maintain concrete at 55°F. for 3 days or longer. Keep concrete warm and moist. Hardened concrete that has been subjected to hard freezing is generally crumbly and usually has to be replaced.
Burns on hands, feet, and exposed skin areas.	Chemical reaction of cement with water forming paste causes material to set-up or harden. Before hardening, concrete can saturate skin and clothes. The basic ingredient, portland cement, is alkaline in nature and therefore caustic.	Prolonged contact with fresh concrete on exposed skin areas may cause rash, allergies, or burns. Avoid such contact. Wash hands with water. Flush exposed skin with water. Get medical attention if reaction persists.
Poor Cylinder Breaks–compression tests are made on cylinders 6-in. in diameter and 12-in. high to measure resistance as expressed in pounds per square inch (psi).	There may be as many as 60 separate and distinct causes of poor or low cylinder breaks–lower than acceptable.	Study, learn and practice good field techniques. Materials available from manufacturers of testing equipment, trade associations, and material suppliers show how this is done.
Stains On Concrete–	Common stains on flatwork include asphalt, oil, iron, copper and aluminum, fire, smoke, ink, mildew, moss, paints, urine, tobacco, and wood stains.	Most stains left on hardened concrete can be removed if proper techniques are used. Chemical or mechanical methods are necessary.

Fig. 20-1 Scaling.

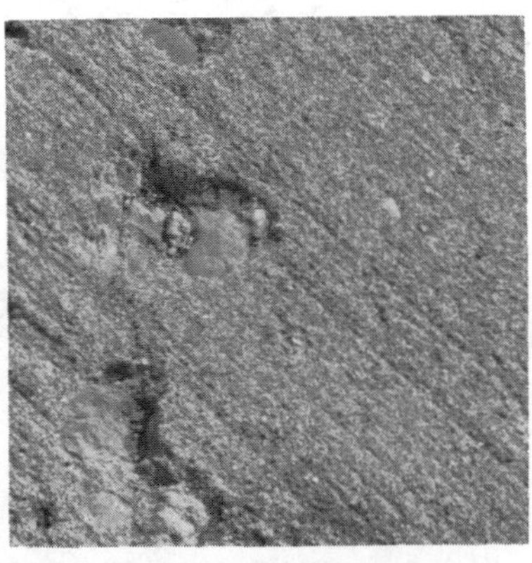
Fig. 20-2 Pop-outs.

Fig. 20-3 Sub-grade failure.

STUDY/DISCUSSION QUESTIONS

1. Can cold weather concreting be accomplished without damaging the properties of concrete?
2. What is the effect of temperature on the rate of hardening?
3. At what temperature should provisions be made for adequate protection during cold weather concreting?
4. Explain three methods for obtaining high-early-strength concrete.
5. Why must the concrete worker be careful when using chemical accelerators?
6. What is the maximum recommended amount of calcium chloride, by weight of cement, that should be used in concrete?
7. Calcium chloride or soluble chloride should never be used under certain conditions. What are these conditions?
8. Is there such a thing as an antifreeze for concrete? Why?
9. Name several things that should be done before placing and finishing concrete in cold weather.
10. Name several things to do after concrete is placed in cold weather.
11. How long should the temperature of concrete be maintained at 70°F.? At 50°F.?
12. What cold weather curing methods are suitable for this area?
13. What precautions are necessary when using salamanders?
14. What may happen to concrete if forms are removed too early?
15. How long should forms remain in place?

UNIT 21

CONCRETE TODAY AND TOMORROW

Look around. No matter where you live or what you do, you will see concrete in use. Formerly the workhorse of the construction industry, concrete is now a material of beauty and versatility. Concrete has come of age.

Concrete is used widely in highway construction, airport installations, home building, for structural and architectural uses, agriculture, and water resources improvements.

PAVING WITH CONCRETE

Concrete serves transportation through roads, streets, expressways, and bridges.

In the Federal-Aid Highway Act of 1956, Congress provided funds for the Interstate Highway System. The standards for this magnificent system are the highest ever set for road construction.

The 41,000-mile national system of interstate and defense highways is the greatest peace-time construction venture in history. Every part of the nation, nearly every city of 50,000 or more population, will be linked by this vast network of highways. Concrete is playing a key part in this program.

The uniformly high skid resistance of concrete in a wet or dry condition, its nighttime visibility, and its even surface are built-in safety features. Tests show that the concrete texture provides the safest pavement surface. This type of skid resistance is built into a concrete pavement with normal finishing procedures.

CONCRETE FOR AIRPORTS

The jet age requires airports capable of handling heavy, high-speed planes. Larger planes and substantial increases in air traffic mean that longer and stronger concrete runways are essential.

Safety is one of the primary reasons for the rapid growth in the use of concrete in airport installations. In any airport, the most vital ground installation is the runway. From the time the wheels of an aircraft first touch the runway until the ship again becomes airborne, the runway must insure a safe, skid-free landing and easy takeoff. The runway must have the strength to sustain tremendous loads.

Safe landings depend to a great extent on good runway visibility. The light color of concrete makes runways highly visible at night. This is a vital factor in achieving economical round-the-clock airport operations.

Concrete runways must be skid resistant and reduce the tendency of an aircraft to veer toward the runway edge. Since runway strips, taxiways, and aprons must sustain very heavy plane weights, the high compressive and beam strength of concrete makes it physically and economically desirable whether the subgrades are of low or high load-bearing capacity.

In addition to structural durability and sound design, facilities for handling jet aircraft must also meet some highly exacting specifications. Concrete is an important factor in meeting some of these requirements. For example, concrete is not affected by the unburned jet fuel that often spills over runways. Concrete does not deteriorate under the terrific heat and blast effect of jets, and it can withstand the punishment of high-pressure jet plane tires.

SOIL-CEMENT PAVING

Soil-cement is a highly compacted mixture of soil or roadway material, portland cement, and water. This mixture forms a strong, durable pavement base. A bituminous surface is placed on the soil-cement base to serve as a wearing surface.

The mixing of soil and portland cement to make a pavement base was a revolutionary idea to many engineers in 1935. However, this mixture proved to be a long-sought method of stabilizing roadway soils to produce a truly satisfactory low-cost pavement.

Soil-cement is now widely used to pave roads, residential streets, airports, and parking areas. Because of its success in these fields, the use of soil-cement expanded to include reservoir linings, dam facings, and subbases for concrete pavement.

BETTER HOMES WITH VERSATILE CONCRETE

Concrete plays a key part in every home. It may serve inconspicuously in footings and basements or dramatically in a beautiful screen wall or curvilinear roof. Concrete and concrete masonry offer the homeowner a material that is versatile and attractive while modest in cost. A durable and firesafe concrete home is both sensible and practical.

The great variety and architectural flexibility of concrete masonry make it usable in homes of any style or size. It can be used in all-concrete construction or in combination with other materials.

Interior masonry walls make the home a safe, interesting place for family and friends to share good companionship.

Walls of concrete block keep the interior environment warm in winter and cool in summer. Little or no maintenance is needed for this durable building material.

Concrete masonry units can be colored by painting or by using integrally colored units. Block textures can be rugged or glass-smooth to the touch. Portland cement paints are especially suitable on exterior surfaces of concrete masonry walls for decorative effects and weatherproofing.

Concrete is used extensively for outdoor living areas, such as in patios, barbecues, swimming pools, and garden walls. These structures are welcome invitations to the open air, whether used by the family alone, a few guests, or large groups of people. Most homeowners prefer concrete sidewalks and driveways because of their durability, cleanliness, and good appearance. Some homeowners use exposed aggregate surfaces, squares, strips, and scored designs for walks and drives to provide design interest.

PRECAST CONCRETE

Before World War II, precasting of structural concrete members was almost unknown in this country. In Europe, however, huge precasting factories turned out the structural components for entire buildings.

Precast buildings are now common in both the United States and Canada. These buildings represent a new technology that goes far beyond fabricating conventional types of

Fig. 21-1 White precast concrete exterior panels showing exposed aggregate.

Fig. 21-2 Cast in place and precast structural elements using white cement.

Fig. 21-3 College dormitory using architectural precast components.

All photographs courtesy Portland Cement Association

structural members on the ground instead of in place. The trend to precast construction has introduced factory precision in design and construction, with resultant economies. The precast units may also be prestressed. This structural flexibility is the result of an intense research and development effort supported by the portland cement industry.

The various uses of precast structural members now span nearly every field of construction. A partial list of uses includes piles and decks for railway and highway bridges, cribbing for retaining walls, railway crossties, floor and roof slabs, wall panels, beams, girders, rigid frames and many specialized products for farm and ranch use.

An entire building can be constructed using precast concrete elements. There is no doubt that precast concrete will play an even greater part in future construction.

STRUCTURAL AND ARCHITECTURAL USES

Structural concrete is characterized by relatively thin sections as compared with the massive concrete sections used in foundations and dams. This type of construction was made possible by the development of reinforced concrete shortly before 1900. In addition, the introduction of techniques such as prestressing, shells, and architectural concrete have improved and expanded the use of concrete as an architectural medium.

Chicago's Marina City (twin 583-ft. towers) is an example of how the development of reinforced and lightweight concrete contributed to high-rise construction.

Concrete is widely used as a casing around structural steelwork for protection against fire. Concrete protects by delaying the flow of heat to the encased steelwork. A 2-inch covering of concrete meets the fire protection requirements of most building codes and fire associations.

Shell design is undoubtedly the most spectacular of all recent developments in concrete. It is now possible to execute almost any shape that the imagination conceives. The TWA terminal at the John F. Kennedy Airport, which suggests a gull in flight, is one of the most striking examples of the fluidity and grace of concrete.

The plasticity of concrete permits unlimited possibilities for unusual ornamental design, flowing lines, smooth curves, and unusual angles. A large variety of colored aggregates, pigments, stains, and paints is available to add color to concrete and create an even more varied appearance.

AGRICULTURAL USES OF CONCRETE

Agriculture is another field in which concrete plays an important role. On the farm, the uses of concrete are almost as varied as the duties of the farmer. Farmers and ranchers use concrete for many purposes because it is an economical solution to their building problems.

Dairy barns, feeding floors, farrowing houses for hogs, poultry houses, silos, and other structures are often constructed of concrete to reduce fire losses, to curb livestock diseases by improving sanitation, and to obtain durable, warm and dry structures. While concrete was used primarily for foundations and floors at first, both reinforced concrete and masonry units are now used extensively in farm construction.

Many farmers fight soil erosion by the use of concrete structures to control both the velocity and flow of runoff water. Concrete is used in modern irrigation systems to store, transport, and distribute water without undue loss by evaporation and seepage. Concrete linings for ditches effectively prevent seepage losses which can average about 40 percent of the water transported in unlined canals. Underground concrete pipelines are also widely used to convey water. Thousands of miles of concrete drain tile are installed yearly to drain wet land and increase its productivity.

WATER RESOURCES

Concrete plays an important part in water control and conservation. As water consumption increases, many cities must transport water from distant natural sources. Other cities impound water in artificial reservoirs. Concrete is used in the construction of dams, spillways, intakes, and pump houses. Concrete pipelines conduct water to homes and cities.

Sewage lines and treatment plants dispose of waste, control disease, and prevent pollution of rivers and lakes. Concrete has been in use for sewer construction since Roman times. Modern concrete sewer pipe varies in diameter from 6 inches to 12 feet.

ENVIRONMENTAL CONCERNS

Solid waste management has become one of the most important concerns of the nation. Counties, towns, regions, and especially larger cities must create plans to deal with the increasing problem of solid waste. Recycling, landfills, incinerators, and futuristic projects that will reclaim liquids, gases, and solids for profitable use are being designed and built. Any proposal must include protection of the air, long and short-term water supply, and the land itself. Versatile concrete and concrete products will take on a vital part in the construction of many of the myriad facilities built to protect our environment.

CONCRETE – THE BUILDING MATERIAL OF THE FUTURE

Concrete is a key material in modern construction. New concrete developments and new products continually improve and extend the use of concrete.

The focal point of research and development is the Portland Cement Association laboratories. Additional research is carried out in several field projects. Examples of such research are the "concrete farms" located at San Pedro, Ca. and Skokie, Ill. At these centers, the performance of concrete posts, slabs, and boxes under various weather conditions is studied. Research for better roads is being conducted on test sections of two-lane concrete highways in several locations. Concrete durability in fresh and salt water is being studied in Massachusetts, New York, Florida, and California. Within the portland cement industry there is a never-ending search for better methods and better products.

But what about future construction with concrete? For an indication, look at the faster, yet safer, system of highways; at dramatic concrete structures; at building research centers. More concrete is used in schools, homes, highways, office buildings, and on the farm than ever before. Concrete is the most widely used building material in the world. Even so, we are just beginning to realize the versatility of this material.

STUDY/DISCUSSION QUESTIONS

1. What qualities of concrete make it the most widely used construction material?

2. Name several general fields of construction where concrete is widely used.
3. What is the Interstate System? When was it authorized?
4. Why is concrete described as the safest pavement available for streets, highways, and airports?
5. Why is concrete especially suited for jet airport facilities?
6. What is soil-cement? What are its uses?
7. What developments have led to the greater architectural and structural flexibility of concrete?
8. What are some of the more important agricultural uses of concrete?
9. In what ways is concrete used for the construction of water supply and sewage disposal facilities?
10. In your opinion, why will concrete be used more and more in the future?
11. Name and discuss some environmental concerns and describe how concrete and concrete products may be used.

APPENDIX

CONCRETE MIX DESIGN – UNIT WEIGHT METHOD

Specifications

JOB LOCATION: ____

CONCRETE TO BE USED FOR: ______________________________________

INFORMATION AND SKETCH OF CONCRETE:

EXPOSURE CONDITIONS:

Mild _______ Severe _______

In Air _____ In Fresh Water ________ In Sea Water or in Contact with Sulfates ____

MAXIMUM W/C RATIO FOR EXPOSURE: _______ lb. water/lb. cement

SPECIFIED STRENGTH: _______ psi at _______ days

MAXIMUM W/C RATIO FOR STRENGTH: _______ lb. water/lb. cement

W/C RATIO TO USE: _______ lb. water/lb. cement

MAXIMUM AGGREGATE SIZE: _______ in.

AIR CONTENT: _______ % ± 1%

RECOMMENDED SLUMP: Minimum: ____ in. Maximum: ____ in. Use: ____ in.

CONCRETE MIX DESIGN – UNIT-WEIGHT METHOD

Date and Calculations for Trial Batch _______

(Saturated, surface-dry aggregates)

Batch size: 10 lb. _____ 20 lb. _____ 50 lb. _____ 100 lb. _____ of cement

Instructions:

a. Complete columns 2, 3, and 4 of table.

b. Fill in items below table.

c. Do calculations 1-5 at bottom.

d. Complete column 5 using N from (5).

(1) Material	(2) Initial wt. (lb.)	(3) Final wt. (lb.)	(4) Wt. Used (Col. 2 - Col. 3) (lb.)	(5) Wt. per Cu. Yd. of Concrete	
				N x Col. 4 (lb./cu. yd.)	Remarks
Cement					Cement Factor
Water					Water Content
Fine Aggregate			(F) =		$\%FA^* = \frac{\text{Wt. of FA}}{\text{Wt. of Agg.}} \times 100$ $F = \frac{F}{F+G} \times 100$
Coarse Aggregate			(G) =		G = ___ x 100 ___ %
Totals				(T)	Total Wt.

*percent fine aggregate of total aggregate

N x T =
___ x ___ = __________

Math check: should equal total wt. above

Amount of Air-entraining Admixture Used: _______ oz.

Measured Slump: ______ in.

Measured Air Content: _______ %

Appearance: Sandy ___ Good ___ Rocky ___

Workability: Excellent ___ Good ___ Fair ___ Poor ___

Wt. of container and concrete (A): ___________ lb.

Wt. of container (B) ___________ lb.

Vol. of container (C) ___________ cu. ft.

(1) Weight of concrete (W) = A - B = ___________ lb.

(2) Unit Weight of concrete (UW) = $\frac{\text{wt. of concrete}}{\text{vol. of container}}$

$= \frac{W}{C} = \text{———} =$ ___________ lb./cu. ft.

(3) Volume of concrete (V) = $\frac{\text{total wt. of material used}}{\text{unit wt. of concrete}}$

$= \frac{T}{UW} = \text{———} =$ ___________ cu. ft.

(4) Yield = vol. of concrete per 100 lb. of cement = $V \times \frac{100}{\text{batch size}}$ = ——— x 100 = ——— cu. ft.

(5) Number of Batches (_____ lb. of cement each) per cu. yd. of

concrete (N) = $\frac{\text{27 cu. ft./cu. yd.}}{\text{vol. of concrete}} = \frac{27}{V} = \frac{27}{?} =$ ___________ batches/cu. yd.

CONCRETE MIX DESIGN – UNIT WEIGHT METHOD

Specifications

JOB LOCATION: ____

CONCRETE TO BE USED FOR: ____________________________________

INFORMATION AND SKETCH OF CONCRETE:

EXPOSURE CONDITIONS:

Mild _______ Severe _______

In Air _____ In Fresh Water ________ In Sea Water or in Contact with Sulfates ____

MAXIMUM W/C RATIO FOR EXPOSURE: _______ lb. water/lb. cement

SPECIFIED STRENGTH: _______ psi at _______ days

MAXIMUM W/C RATIO FOR STRENGTH: _______ lb. water/lb. cement

W/C RATIO TO USE: _______ lb. water/lb. cement

MAXIMUM AGGREGATE SIZE: _______ in.

AIR CONTENT: _______ % ± 1%

RECOMMENDED SLUMP: Minimum: ____ in. Maximum: ____ in. Use: ____ in.

CONCRETE MIX DESIGN – UNIT-WEIGHT METHOD

Date and Calculations for Trial Batch _______

(Saturated, surface-dry aggregates)

Batch size: 10 lb. _____ 20 lb. _____ 50 lb. _____ 100 lb. _____ of cement

Instructions:

a. Complete columns 2, 3, and 4 of table.

b. Fill in items below table.

c. Do calculations 1-5 at bottom.

d. Complete column 5 using N from (5).

(1) Material	(2) Initial wt. (lb.)	(3) Final wt. (lb.)	(4) Wt. Used (Col. 2 - Col. 3) (lb.)	(5) Wt. per Cu. Yd. of Concrete	
				N x Col. 4 (lb./cu. yd.)	Remarks
Cement					Cement Factor
Water					Water Content
Fine Aggregate			(F) =		$\%FA^* = \frac{\text{Wt. of FA}}{\text{Wt. of Agg.}} \times 100$ $F = \frac{F}{F+G} \times 100$
Coarse Aggregate			(G) =		G = ___ x 100 ___ %
Totals				(T)	Total Wt.
					Math check: should equal total wt. above

*percent fine aggregate of total aggregate

N x T =
___ x ___ = _________

Amount of Air-entraining Admixture Used: _______ oz.

Measured Slump: ______ in.

Measured Air Content: _______ %

Appearance: Sandy ___ Good ___ Rocky ___

Workability: Excellent ___ Good ___ Fair ___ Poor ___

Wt. of container and concrete (A): ___________ lb.

Wt. of container (B) ___________ lb.

Vol. of container (C) ___________ cu. ft.

(1) Weight of concrete (W) = A - B = ___________ lb.

(2) Unit Weight of concrete (UW) $= \frac{\text{wt. of concrete}}{\text{vol. of container}}$

$= \frac{W}{C} =$ ______ = ___________ lb./cu. ft.

(3) Volume of concrete (V) $= \frac{\text{total wt. of material used}}{\text{unit wt. of concrete}}$

$= \frac{T}{UW} =$ ______ = ___________ cu. ft.

(4) Yield = vol. of concrete per 100 lb. of cement $= V \times \frac{100}{\text{batch size}}$ = ______ x 100 = ______ cu. ft.

(5) Number of Batches (_____ lb. of cement each) per cu. yd. of

concrete (N) $= \frac{\text{27 cu. ft./cu. yd.}}{\text{vol. of concrete}} = \frac{27}{V} = \frac{27}{?} =$ ___________ batches/cu. yd.

CONCRETE MIX DESIGN – UNIT WEIGHT METHOD

Specifications

JOB LOCATION: ____

CONCRETE TO BE USED FOR: ____________________________________

INFORMATION AND SKETCH OF CONCRETE:

EXPOSURE CONDITIONS:

Mild _______ Severe _______

In Air _____ In Fresh Water ________ In Sea Water or in Contact with Sulfates ____

MAXIMUM W/C RATIO FOR EXPOSURE: _______ lb. water/lb. cement

SPECIFIED STRENGTH: _______ psi at _______ days

MAXIMUM W/C RATIO FOR STRENGTH: _______ lb. water/lb. cement

W/C RATIO TO USE: _______ lb. water/lb. cement

MAXIMUM AGGREGATE SIZE: _______ in.

AIR CONTENT: _______ % ± 1%

RECOMMENDED SLUMP: Minimum: ____ in. Maximum: ____ in. Use: ____ in.

CONCRETE MIX DESIGN – UNIT-WEIGHT METHOD

Date and Calculations for Trial Batch _______

(Saturated, surface-dry aggregates)

Batch size: 10 lb. _____ 20 lb. _____ 50 lb. _____ 100 lb. _____ of cement

Instructions:

a. Complete columns 2, 3, and 4 of table.

b. Fill in items below table.

c. Do calculations 1-5 at bottom.

d. Complete column 5 using N from (5).

(1) Material	(2) Initial wt. (lb.)	(3) Final wt. (lb.)	(4) Wt. Used (Col. 2 - Col. 3) (lb.)	(5) Wt. per Cu. Yd. of Concrete: N x Col. 4 (lb./cu. yd.)	(5) Wt. per Cu. Yd. of Concrete: Remarks
Cement					Cement Factor
Water					Water Content
Fine Aggregate			(F) =		$\%FA^* = \frac{\text{Wt. of FA}}{\text{Wt. of Agg.}} \times 100$ $F = \frac{F}{F+G} \times 100$
Coarse Aggregate			(G) =		G = ___ x 100 ___ %
Totals				(T)	Total Wt.
*percent fine aggregate of total aggregate			N x T = ___ x ___ = _________		Math check: should equal total wt. above

Amount of Air-entraining Admixture Used: _______ oz. Measured Slump: ______ in.

Measured Air Content: _______ %

Appearance: Sandy ___ Good ___ Rocky ___

Workability: Excellent ___ Good ___ Fair ___ Poor ___

Wt. of container and concrete (A): ___________ lb.

Wt. of container (B) ___________ lb.

Vol. of container (C) ___________ cu. ft.

(1) Weight of concrete (W) = A - B = ___________ lb.

(2) Unit Weight of concrete (UW) = $\frac{\text{wt. of concrete}}{\text{vol. of container}}$

$= \frac{W}{C} =$ ______ = ___________ lb./cu. ft.

(3) Volume of concrete (V) = $\frac{\text{total wt. of material used}}{\text{unit wt. of concrete}}$

$= \frac{T}{UW} =$ ______ = ___________ cu. ft.

(4) Yield = vol. of concrete per 100 lb. of cement = V x $\frac{100}{\text{batch size}}$ = ______ x 100 = ______ cu. ft.

(5) Number of Batches (_____ lb. of cement each) per cu. yd. of

concrete (N) = $\frac{\text{27 cu. ft./cu. yd.}}{\text{vol. of concrete}} = \frac{27}{V} = \frac{27}{?} =$ ___________ batches/cu. yd.

CONCRETE MIX DESIGN – UNIT WEIGHT METHOD

Specifications

JOB LOCATION: ____

CONCRETE TO BE USED FOR: ______________________________________

INFORMATION AND SKETCH OF CONCRETE:

EXPOSURE CONDITIONS:

Mild _______ Severe _______

In Air _____ In Fresh Water _______ In Sea Water or in Contact with Sulfates ____

MAXIMUM W/C RATIO FOR EXPOSURE: _______ lb. water/lb. cement

SPECIFIED STRENGTH: _______ psi at _______ days

MAXIMUM W/C RATIO FOR STRENGTH: _______ lb. water/lb. cement

W/C RATIO TO USE: _______ lb. water/lb. cement

MAXIMUM AGGREGATE SIZE: _______ in.

AIR CONTENT: _______ % ± 1%

RECOMMENDED SLUMP: Minimum: ____ in. Maximum: ____ in. Use: ____ in.

CONCRETE MIX DESIGN – UNIT-WEIGHT METHOD

Date and Calculations for Trial Batch _______

(Saturated, surface-dry aggregates)

Batch size: 10 lb. _____ 20 lb. _____ 50 lb. _____ 100 lb. _____ of cement

Instructions:

a. Complete columns 2, 3, and 4 of table.
b. Fill in items below table.
c. Do calculations 1-5 at bottom.
d. Complete column 5 using N from (5).

(1) Material	(2) Initial wt. (lb.)	(3) Final wt. (lb.)	(4) Wt. Used (Col. 2 - Col. 3) (lb.)	(5) Wt. per Cu. Yd. of Concrete: N x Col. 4 (lb./cu. yd.)	(5) Wt. per Cu. Yd. of Concrete: Remarks
Cement					Cement Factor
Water					Water Content
Fine Aggregate			(F) =		$\%FA^* = \frac{\text{Wt. of FA}}{\text{Wt. of Agg.}} \times 100$ $F = \frac{F}{F+G} \times 100$
Coarse Aggregate			(G) =		G = ___ x 100 ___ %
Totals				(T)	Total Wt.
					Math check: should equal total wt. above

*percent fine aggregate of total aggregate

N x T =
___ x ___ = _________

Amount of Air-entraining Admixture Used: _______ oz.

Measured Slump: ______ in.

Measured Air Content: _______ %

Appearance: Sandy ___ Good ___ Rocky ___

Workability: Excellent ___ Good ___ Fair ___ Poor ___

Wt. of container and concrete (A): ___________ lb.

Wt. of container (B) ___________ lb.

Vol. of container (C) ___________ cu. ft.

(1) Weight of concrete (W) = A - B = ___________ lb.

(2) Unit Weight of concrete (UW) $= \frac{\text{wt. of concrete}}{\text{vol. of container}}$

$= \frac{W}{C} = \text{———} =$ ___________ lb./cu. ft.

(3) Volume of concrete (V) $= \frac{\text{total wt. of material used}}{\text{unit wt. of concrete}}$

$= \frac{T}{UW} = \text{———} =$ ___________ cu. ft.

(4) Yield = vol. of concrete per 100 lb. of cement $= V \times \frac{100}{\text{batch size}}$ = ——— x 100 = ——— cu. ft.

(5) Number of Batches (_____ lb. of cement each) per cu. yd. of

concrete (N) $= \frac{\text{27 cu. ft./cu. yd.}}{\text{vol. of concrete}} = \frac{27}{V} = \frac{27}{?} =$ ___________ batches/cu. yd.

Conversion Tables, Concrete Mix Design by Weight

CEMENT CONTENT per cubic yard		
Barrels	Bags	Pounds
1	4	376
	4.5	423
1.25	5	470
	5.5	517
1.5	6	564
	6.5	611
1.75	7	658
	7.5	705
2	8	752
	8.5	799
2.25	9	846

Tonnage Conversion

1 ton = 5.32 bbl. (376 lb. per bbl.)

= 21.28 bags (94 lb. per bag)

lb./cu. yd. to kg/m^3: multiply by 0.4933

WATER-CEMENT RATIO*	
Gal./bag	Weight Ratio
4	0.36
4.5	0.40
5	0.44
5.5	0.49
6	0.53
6.5	0.58
7	0.62

*gal./bag to weight ratio
Multiply by 0.0888

Metric Equivalents

CEMENT CONTENT	
lb./cu. yd.	kg/m^3
376	223
423	251
470	279
517	307
564	335
611	362
658	390
705	418
752	446
799	474
846	502

CEMENT CONTENT	
kg/m^3	lb./cu. yd.
200	337
225	379
250	421
275	464
300	506
325	548
350	590
375	632
400	674
425	716
450	758
475	801
500	843

For:	Multiply by:
Cement content lb./cu. yd. to kg/m^3	0.5933
Water-cement ratio gal./bag to weight ratio	0.0888

INDEX

Index